The
Drill Press Book
Including 80 Jigs & Accessories You Can Make

The
Drill Press Book
Including 80 Jigs & Accessories You Can Make

R. J. De Cristoforo

TAB BOOKS
Blue Ridge Summit, PA

FIRST EDITION
THIRD PRINTING

© 1991 by **TAB Books**.
TAB Books is a division of McGraw-Hill, Inc.

Library of Congress Cataloging-in-Publication Data

De Cristoforo, R. J.
    The drill press book: Including 80 jigs & accessories you can make
/ by R. J. De Cristoforo.
      p.   cm.
    Includes index.
    ISBN 0-8306-7609-0   ISBN 0-8306-3609-9 (pbk.)
    1. Drill presses. I. Title.
TJ1260.D437   1991   90-21678
621.9′52—dc20   CIP

TAB Books offers software for sale. For information and a catalog, please contact TAB Software Department, Blue Ridge Summit, PA 17294-0850.

Acquisitions Editor: Kimberly Tabor
Book Editor: Cherie R. Blazer
Director of Production: Katherine G. Brown
Book Design: Jaclyn J. Boone
Cover photograph courtesy of Delta International Machinery Corporation   TAB1

Contents

Introduction

Excluding a sawing tool, the drill press, whether a bench or floor model, is the most important power tool in a woodworking shop. After project components are cut to size, they must be drilled for fasteners, organized for connections like dowel joints and mortises, possibly embellished with decorative edges or incised details, and sanded. The drill press is available for all these chores. It can also be used for fluting and reeding of spindles, forming discs, producing wheels for toys and other projects, sanding or routing multiple pieces precisely, making joint dowels or screw concealment plugs. The list of possible functions extends to the point of fascination.

Some of the machine's applications are made possible by commercial accessories, but there are many jigs and accessories the user can make with little effort and at minimum cost. This book offers detailed instructions for homemaking many units that do as well as costly commercial ones, and others that are just not available otherwise.

The machine is a simple concept: a motor-driven vertical spindle with an assortment of speeds (RPM), and a chuck at its free end for gripping drill bits and other cutters. A basic welcome advantage is that drilled holes will automatically be square to the surface of the workpiece, a factor that requires operational skill when attempted with portable tools.

Space requirements are negligible. A floor model occupies a few square feet, and it becomes mobile when its base is secured to a suitable pallet that is equipped with lockable casters. The term "bench model" doesn't mean attachment to a workbench where the drill press can interfere with other chores. It is usually mounted on its own cabinet-type stand.

Drill presses are available in a variety of sizes and prices to suit any shop plan and budget. A floor model will cost more than a bench model but in many cases a manufacturer's distinction between the two versions is only in the length of the column.

There is much enjoyment in using power tools, but also potential hazards. The drill press is no exception, even though it is often touted as one of the safer machines. No machine is "safe." No matter how sophisticated or practical, it is a brainless piece of equipment that does your bidding, disinterested in what is under the cutter or on the worktable, or whether the work is secured or free to spin with the cutter.

Approach the training session with the understanding that you are in control. You are accountable for misuse, abuse, and ignoring the correct procedures and safety rules that apply to shop work in general as well as the tool. Remember the adage "familiarity breeds contempt." In relation to power tools, don't become overconfident as you gain knowledge. Expertise is not a synonym for safety. There are many professionals with years of experience who attest to that.

Get to know the tool before using it by studying the owner's manual; it's the bible for that particular unit. Obey operating procedures. Take the time to do "dry runs," especially when the operation is foreign to you. That is, follow the steps with the machine turned off. This way you can preview what will occur when you situate the workpiece, and find the best location for your hands and body.

The drill press does not have a guard. In general, safety calls for work security and sensible hand positions. With some applications, like routing and shaping, a safety shield makes sense. The items are not available for purchase but the book offers designs so you can make them for your personal use. I use them, so should you.

Consider these two basic rules when using the drill press:

Measure twice before drilling and think twice before turning on the tool.

Chapter **1**

The drill press

Dozens of drill press models are available but, except for particular features, they all function the same way and have similar components (FIG. 1-1). The products currently available are too numerous to mention here, but TABLE 1-1 lists basic specifications of example units.

The machine consists of a tubular, vertical column that is secured to a substantial base and that supports a vertically adjustable worktable, preferably one that tilts. A fixed *headstock* at the top of the column houses the tool's working mechanism. A good drill press will have a surface-ground base so it can substitute for the standard table when a chore requires more depth under the chuck than is available normally.

An exception to this general configuration is the Shopsmith multipurpose machine when it is set in its drill press mode (FIG. 1-2). Here, the headstock that is the power source of the tool moves on twin, tubular ways so its height posture is flexible. The table is also movable vertically and, because it is the sawing component, its rip fence and miter gauge come into play as guides and jigs for many drill press operations.

The difference between standard floor and bench models is apparent in FIGS. 1-3 and 1-4. The manufacturer often supplies identical essentials so the only difference between the units, other than cost, is the length of the column. This length relates to the maximum height or length of work that can be situated between the chuck and table, or base. A bench unit can be secured to a fixed surface or to a workbench, but does not have to be. Actually, it is more practical to mount the tool on a special stand, as in FIG. 1-5. The design that is suggested is basic but it can be closed at the back and fitted with shelves and a door so it can provide accessory storage as well as tool support.

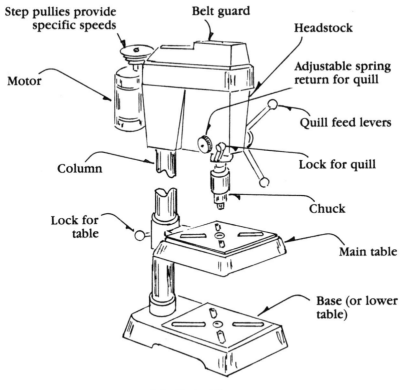

Step pullies provide
specific speeds

Belt guard

Headstock

Adjustable spring
return for quill

Motor

Quill feed levers

Column

Lock for quill

Lock for
table

Chuck

Main table

Base (or lower
table)

1-1 Basic drill press.

Both a stand-mounted bench model and a floor model use about the same amount of shop space. A floor model can remain in a permanent location (many operators bolt it to the floor), or it can be made mobile by mounting it on a pallet (FIG. 1-6). When so mounted, it is easy to move the tool from the work area when it is not in use. Be sure, when selecting casters for the stand or the pallet, that they have a locking device. You don't want the tool to move when it is being used.

THE SPINDLE

A vital component of a drill press is the spindle that is situated vertically in the headstock. The spindle rotates in bearings that are press-fitted in a sleeve—technically, the *quill*. Part of the spindle's length is splined (grooved) to mate with matching configurations in the driven pulley. This design permits vertical spindle movement while both it and the pulley are rotating.

Some non-drilling operations such as routing and shaping involve side thrust, so the question of whether the applications can harm spindle bearings often arises. Actually, it's difficult to find a modern machine that isn't equipped with heavy-duty ball bearings that will stand up to radial loading. A machine equipped with sleeve bearings is another story. In such a case, radial stresses should probably be avoided or, at least, minimized. Relevant information will be found in the owner's manual. If not, check with the supplier or manufacturer.

Table 1-1 Drill Press Specifications

BRAND	MODEL	SIZE	CHUCK	POWER	SPEEDS Belt	SPEEDS Variable	TYPE Bench	TYPE Floor	TABLE	QUILL STROKE
Delta	17-441	17	1/2	1HP	(5)700-4250		X	X	11 × 14	5
	15-321	15	1/2	3/4 HP	(12)250-3000	450-4700	X	X	11 × 14	6
	17-900	16 1/2	5/8	3/4 HP				X	12 × 12	3.35
Sears	213852	15	5/8	1/2 HP	(12)300-4600			X	12 × 12	3
	213830	10	5/8	1/4 HP	(4)480-3000		X		8 × 8	2
Black & Decker	1782	16 1/2	5/8	1/2 HP	(12)250-3000			X	12 × 18 1/4	4 21/32
	9400	8	1/2	4.5 amp	(5)620-3100		X		6 3/8 square	2
Bridgewood	1012B	10	1/2	1/2 HP	(5)540-3600		X		7 1/2 square	2 1/2
	1412F	14	1/2	1/2 HP	(5)460-2420			X	12 diameter	3 3/8
Jet	708650	8	1/2	1/6 HP	(5)620-3100		X		6 3/8 diameter	2
	35406	17	5/8	1/2 HP	(16)250-3650		X		13 3/4 diameter	4 3/8
AMT	4560	12 1/2	1/2	3/4 HP	(12)250-3100			X	8 diameter	3 1/4
	4590	12 1/2	1/2	3/4 HP	(12)250-3100		X		8 diameter	3 1/4

I-2 The Shopsmith multipurpose tool in its drill press mode. The headstock and the table move on twin, tubular ways. Note the built-in speed changer and the availability of rip fence and miter gauge for drilling operations.

When doing operations like routing or shaping, it's necessary to use special chucks in place of the standard one. Pertinent information will be offered in the chapters dealing with those subjects. Some machines are designed for inter-changeable spindles so special ones can be selected for, among other things, operations involving side thrust. Such units approach industrial specifications—and

1-3 A floor model drill press has impressive capacity between the chuck and base.

1-4 Bench model drill press.

price—but the work can be done on less sophisticated products by relying on economical adapters.

The quill, together with the spindle, is moved downward for drilling and other operations by means of a lever or levers that are located on the right side of the headstock. When the feed lever is released, the quill automatically returns to its neutral position, retracted by a coil spring that is adjustable for tension. The quill should return easily and quietly. Too much spring tension will return the quill with a shock, and might even cause the feed levers to bang against your hand.

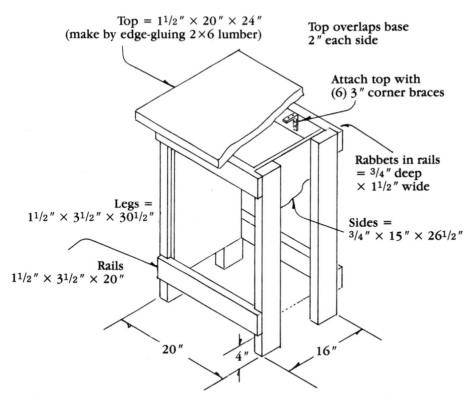

Top = 1¹/₂" × 20" × 24"
(make by edge-gluing 2×6 lumber)

Top overlaps base
2" each side

Attach top with
(6) 3" corner braces

Rabbets in rails
= ³/₄" deep
× 1¹/₂" wide

Legs =
1¹/₂" × 3¹/₂" × 30¹/₂"

Sides =
³/₄" × 15" × 26¹/₂"

Rails
1¹/₂" × 3¹/₂" × 20"

20"

4"

16"

1-5   It's a good idea to mount a bench model drill press on its own stand. This design will be suitable for storage if a back, shelves, and a door are added.

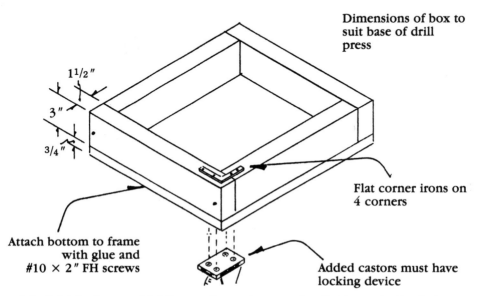

Dimensions of box to
suit base of drill
press

1¹/₂"

3"

³/₄"

Flat corner irons on
4 corners

Attach bottom to frame
with glue and
#10 × 2" FH screws

Added castors must have
locking device

1-6   Pallet for floor model drill press. Be sure to use casters that have a locking feature.

QUILL CONTROLS

Quill travel (FIG. 1-7) is the maximum distance the quill can be moved downward by the feed levers, thus the deepest hole that can be drilled without relying on special techniques.

A depth-stop rod (FIG. 1-8) is used to limit the quill's extension. A typical application is drilling a blind or stopped hole—one that doesn't pass completely through the workpiece. The lock nut is secured on the rod so its distance from the casting on which the rod sits equals the depth of the hole you need. This is an especially useful feature when a project requires multiple holes of similar depth. Don't, for any reason, work without the lock nut on the stop rod. This would allow the quill to extend so far that it could fall from the headstock.

There are many operations, typically routing, shaping, and drum sanding, that require the quill to be locked in an extended position. This is accomplished with a lever or similar device (FIG. 1-9). The device can also be used to hold the quill at a particular extension while you set the stop rod for drilling blind holes.

CAPACITIES

The size of a drill press is designated as twice the distance from the column to the vertical centerline of the spindle. For example, a 17-inch drill press will be able to

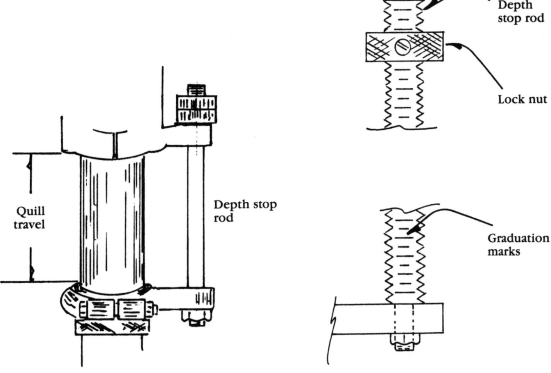

1-7 Quill travel is the maximum distance the quill can be extended.

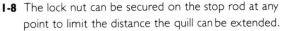

1-8 The lock nut can be secured on the stop rod at any point to limit the distance the quill can be extended.

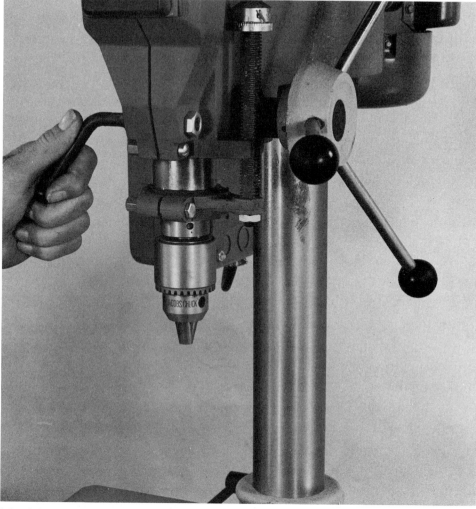

I-9 A lever or similar device is used to lock the quill at an extended position. The feature is needed when jobs require that the cutter be held at a particular point above the worktable.

drill at the center of stock that is 17 inches wide (FIG. 1-10). A second capacity, not always revealed, is the maximum distance from the chuck to the table or base (FIG. 1-11). The floor model drill press stands as champion here, but the astute bench model operator can compete by swinging the tool's headstock around so the spindle will extend beyond the edge of whatever the tool is mounted on. Then, work length can extend from the chuck to the floor.

THE CHUCK

The standard tool holder for a drill press is a three-jaw chuck and matching key (FIG. 1-12). The barrel of the chuck can be turned by hand to spread the jaws and even to bring the jaws in contact with the shank of the tool, but the final tighten-

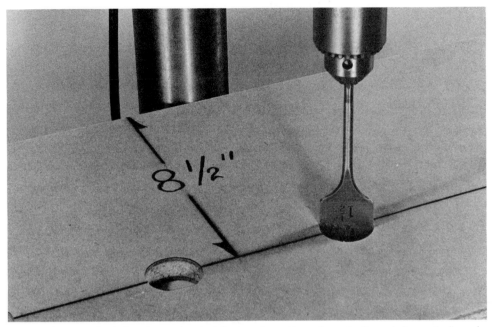

1-10 Drill press size is listed as twice the distance from the column to the spindle centerline. In this case, the machine would be a 17-inch unit.

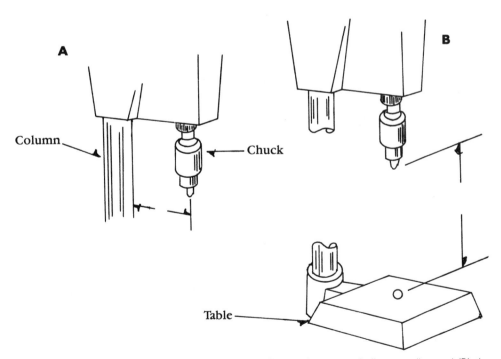

1-11 Drill press capacities: (A) Twice the distance from column to spindle centerline and (B) the distance from the chuck to the lowest table position (or base).

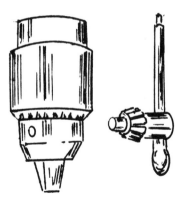

I-I2   Typical three-jaw chuck and key.

ing or the release of the tool must be done with the key. If the jaws are not tight on the tool, the chuck will spin around the shank, abrading it and not accomplishing anything else.

Typical chuck sizes range from 1/4 to 5/8 inch. This determines the diameter of the shank the chuck can grip, but not necessarily the tool's cutting diameter. For example, many drill bits for large holes will have 1/4-inch shanks. Thus a "small" chuck can accommodate a "large" drill bit. A more critical factor in this area is the power of the machine. If cutting stalls or the motor overheats or slows up considerably, you'll know the operation won't work, even though the chuck can grip the bit.

The capacity of the chuck is rated from the smallest to the largest shank it will take. A quality unit will start at zero; it will securely grip the smallest bit you have. Others might not be capable of holding anything smaller than 1/32 or 1/16 inch.

Chucks mount on the end of the spindle. The method used depends on the design of the machine. Some are screwed on and then secured with a screw that passes through the chuck into the spindle. A more sophisticated and usually more precise attachment involves a chuck with a tapered hole that matches the end shape of the spindle (FIG. 1-13). In addition, the chuck and spindle are threaded so that tightening the chuck's collar seats the matching tapers exactly. Be sure to maintain chuck and spindle in pristine condition.

SPEEDS

To accomplish various applications efficiently and safely, the drill press must be organized for different speeds (spindle RPM). This is usually accomplished by a cone-shaped step pulley at the top of the spindle that connects by V-belt to a matching pulley on the motor shaft (FIG. 1-14). The number of speeds available is determined by the number of steps on the pulleys. Four-speed pulleys providing a speed range of about 400 to 4000 RPM are not uncommon, although this is not a standard specification. TABLE 1-1 shows that, while four or five speeds seem to be a minimum, a drill press can be designed to provide as many as twelve or sixteen.

It's possible to establish additional speeds by installing an intermediate or *jackshaft* pulley between the standard pulleys (FIG. 1-15). This can't be done hap-

I-13 Tapered spindle accepts a matching tapered hole in the chuck. The chuck collar is threaded on the spindle to draw the chuck tightly onto the spindle.

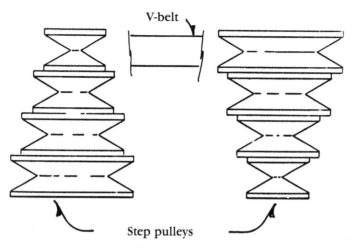

1-14 Common speed changing method involves matching step pulleys—one on the spindle and the other on the motor shaft.

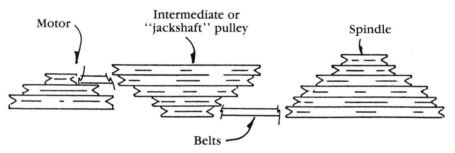

1-15 Additional speeds can be obtained by installing a jackshaft pulley.

hazardly or on all machines. Check the owner's manual or with the manufacturer to determine if the idea is feasible.

Changing the V-belt position can be a finger-nipping procedure unless you work carefully. Most machines provide for tilting the motor or moving it forward to release belt tension (FIGS. 1-16 and 1-17). After the belt change, the motor is returned to its correct position and secured by a lever or some other means that bears against motor mount rods (FIG. 1-18).

The degree of belt tension is important: too loose, and it will slip on the pulleys, too tight and there will be excessive strain on both the motor shaft and the spindle. Usually, if the belt can be flexed about 1/8 inch with light finger pressure the tension is correct. Check machine instructions to be sure.

Some machines provide for infinitely variable speeds with an installed speed changer (FIG. 1-19). Turning the dial expands or contracts a movable sheave on the spindle—thus, in effect, changing the diameter of the driven pulley. The concept is a great convenience because it eliminates belt changing. It also has operational advantages: You can command any speed between minimum and maximum and are not confined to the specific speeds of step pulleys.

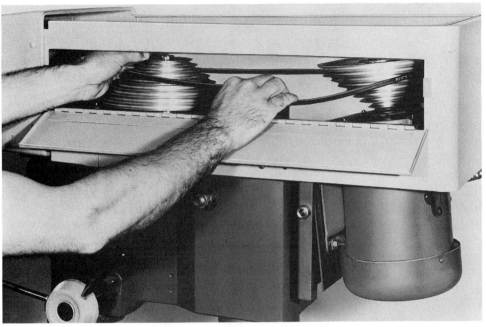

1-16 Belt changing on some machines is easier because the motor can be tilted toward the headstock.

1-17 Another idea for belt changing: The motor, on its mount, can be moved forward, thus changing the distance between pulleys. The motor is moved back to its original position to reestablish belt tension.

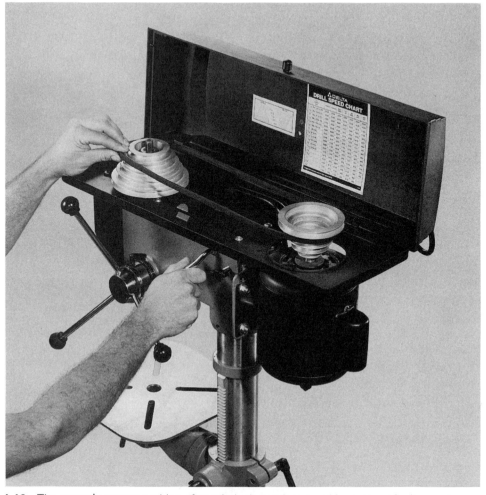

1-18 The motor's correct position after a belt change is secured by means of a lever or some other locking device. Note the speed chart installed in the belt guard.

The general rule for speed selection is to go more slowly with larger tools. However, this is not always the most efficient way to work with particular cutters. For example, a spade bit, even one as large as 1¹/₂ inches, functions best at a speed that would be impractical with a different but similar-size tool—such as a fly cutter or hole saw. These considerations will be explained as we go along.

Some machines provide a chart that suggests an efficient and safe speed for various operations (FIG. 1-20). Most times, especially when in doubt, it's wise to start at a slow speed and go higher until the cutter is working efficiently. The tool must cut steadily. Allowing it just to rub won't accomplish much and will dull cutting edges.

Speed considerations will be discussed as they apply to various applications in the following chapters.

I-19 Built-in, manual speed changer provides infinitely variable speed changes from 450 to 4700 RPM. Make changes only when the machine is running.

I-20 Sears unit has a chart that suggests the correct speed for various applications.

TABLES

The worktable moves vertically on the column and can be locked at any point between the headstock and base. It can be swung aside to free the area between chuck and base and it should be tiltable so it can be adjusted for, among other things, angular drilling. Two basic types are available.

The first type of table, conceived for metal working, has both diagonal and parallel blind T-slots to facilitate clamping of jigs and fixtures. It also has a border trough to catch lubricants that are often required when doing metal drilling (FIG. 1-21). The second and more common type is especially suitable for woodworking. It has a less sophisticated slot design, but adds a center hole and side ledges that facilitate clamping of workpieces (FIG. 1-22). The unit shown has an index pin to accurately locate the table in its neutral position and in 90-degree left or right tilt postures. A scale is used for intermediate tilt settings, but for the sake of accuracy an adjustable bevel should be used to check the angle between cutter and table before work is done.

ADDITIONAL CONSIDERATIONS

According to standard practice, the height of the worktable is adjusted manually. This isn't difficult to do, but some units are more convenient than others. Delta

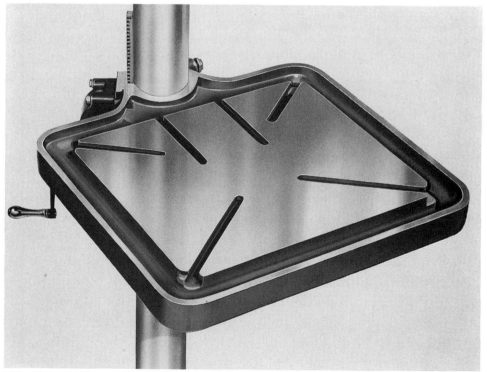

1-21 This table is good for metal working, with specially designed, blind T-slots and a border trough to catch lubricants.

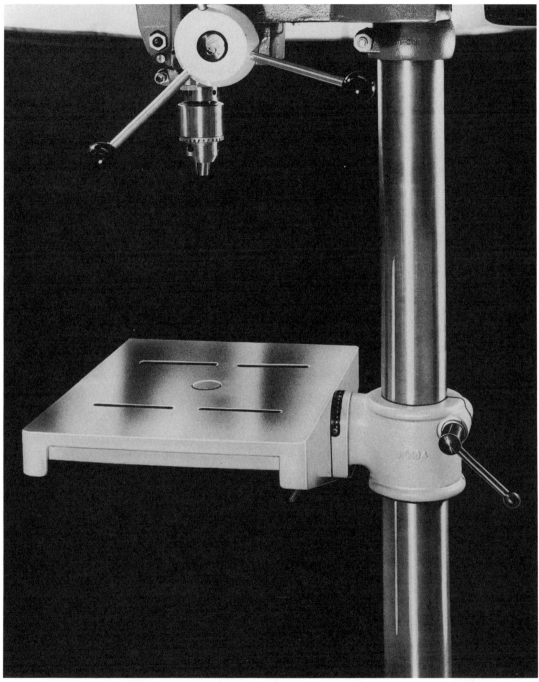

1-22 This table is especially good for woodworking. It has a center hole and slots that penetrate the table, plus clamp ledges that facilitate securing workpieces.

1-23 A table raising and lowering mechanism is available for some machines either as standard or accessory equipment.

provides as standard equipment, or as an accessory for some of its models, a table raising and lowering mechanism that facilitates positioning the drill press table to suit different size jobs (FIG. 1-23). Sears provides a similar design as standard equipment on some of its floor and bench model machines (FIG. 1-24).

A foot feed (FIG. 1-25) is a boon for work that requires repeated extending and retraction of the spindle. This type of accessory is usually designed for columns of a particular length and they are not interchangeable from brand to brand. However, Chapter 13 gives instructions on how you can make a foot feed that will be just right for the machine you own.

NEW CONCEPTS

Electronics are having an impact on power tools. While there seems little evidence at this writing that this applies to large home-type drill presses, a few manu-

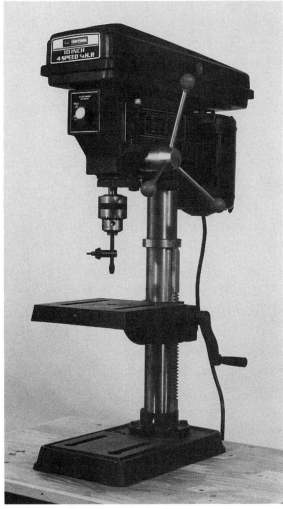

I-24 Sears supplies a crank device for table height adjustments on both bench and floor model machines.

I-25 A foot feed lever is convenient for many operations. The accessory is not available for all machines, but you can make one.

facturers seem willing to open the door. Black & Decker's small bench model employs a microcomputer that responds to various commands (FIG. 1-26). A display panel will show preselected drilling RPM and even the position of the drilling head in inches. There are speed selection keys and you can read on the display panel the depth to which you are drilling. It takes a little practice to fully utilize the concept, but the owner's manual is detailed and leads you step-by-step through various functions.

DRILL STANDS

Many workers who are not inclined to add a full-size drill press to their shop equipment have discovered that they can do a reasonable amount of drill press

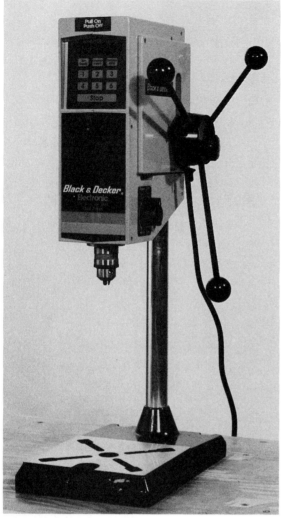

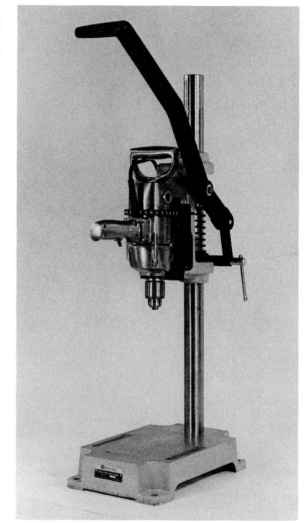

1-26 This small drill press has a microcomputer and a display panel. Buttons are used for speed settings; the display reveals the amount of quill extension.

1-27 Delta's drill stand is a sturdy unit that will accommodate heavy, 1/2-inch drills. The drill mount is spring-loaded so it has an automatic retraction feature.

work by combining a portable drill with a special stand. There are quite a few of these accessories available. While some of them can accommodate various portable drill sizes and models, others are suited only to the manufacturer's own products. The portable drill/stand units are not suitable for all the applications that will be explained in this book, but for basic drilling and chores like drum sanding and so on, they can do respectable work. Examples are shown in FIGS. 1-27 and 1-28.

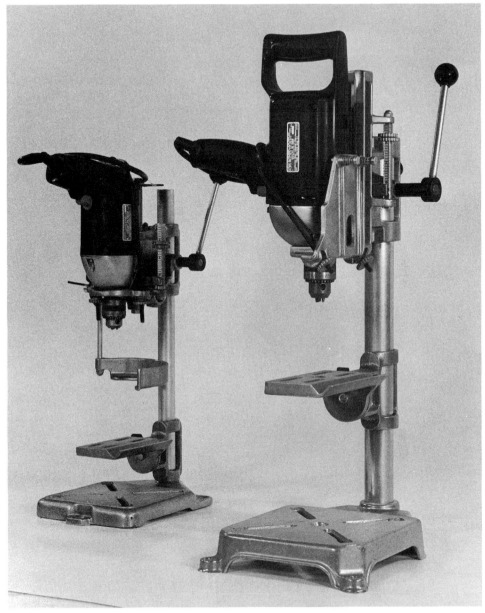

I-28 Sears offers various drill stand designs that include worktables and bases just like a regular drill press.

Chapter **2**

Drilling tools

*B*oring, *drilling*, and *bits* are terms continually used when working with the drill press. Technically, you bore in wood and drill in metal. Common usage has made the words synonyms; they both refer to forming holes. The bit refers to the tool itself.

There is a difference between drilling tools for wood and those especially suitable for metal, although there is overlap in some areas. Hole saws and fly cutters, for example, are usable on both materials. Twist drills are perfect for metal drilling but are commonly used to drill wood, even though they don't cut as cleanly as bits expressly designed for the purpose.

Drilling tools have individual characteristics, as you can see by the cross-section views of blind holes in FIG. 2-1. On through holes, twist drills and spade bits do not break through the material as cleanly as brad point bits. Forstner bits finish a hole in good style and will produce clean, true, flat-bottom holes.

Bits with screw points (FIG. 2-2) work well in a hand brace but should not be used under power. The purpose of the screw is to pull the bit into the work. This is fine when working by hand because it minimizes how hard you must bear down on the tool, but under power the screw will attempt to control the rate of penetration. This would work if the RPM could be adjusted in relation to the lead of the screw, but it's feasible only under rigidly controlled conditions. That's why it's more efficient and safer to use bits that have points, so work penetration is controlled by the operator, not the bit.

Workers who have a collection of screw-point bits and wish to use them under power can modify them by filing off part or all of the threads to form a point. This must be done carefully so the point will be on the longitudinal centerline of the tool.

Many drilling tools have *flutes* spiraling up the shank from the cutting edges. These are channels up which waste chips can travel to clear the hole, preventing

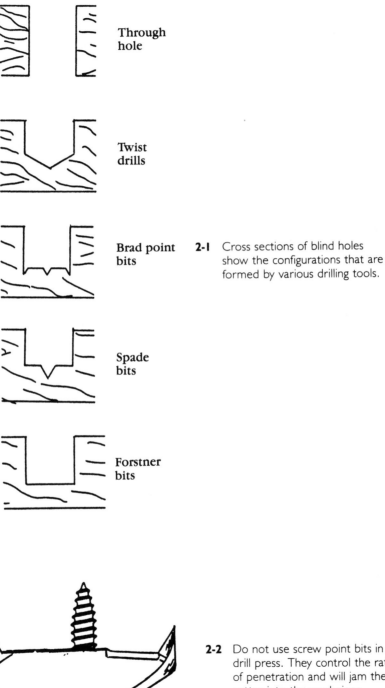

Through
hole

Twist
drills

Brad point
bits

Spade
bits

Forstner
bits

2-1 Cross sections of blind holes
show the configurations that are
formed by various drilling tools.

2-2 Do not use screw point bits in a
drill press. They control the rate
of penetration and will jam the
cutter into the workpiece.

waste build-up that would cause overheating and inaccurate work. The bit should be frequently retracted; waste must not accumulate in the hole.

TWIST DRILLS

Twist drills have a very high efficiency rating. Technically, this is based on the length of cutting edges in proportion to the back-up metal that supports them. It's a long-lived tool and will keep cutting even when abused. However, always respect the tool and its limits if you want optimum results.

Although twist drills are used extensively, they really don't do the best job in wood. The common twist drill will be better suited for wood drilling if its cutting angle is about 40 degrees, as opposed to the 59-to 60-degree angle that is better

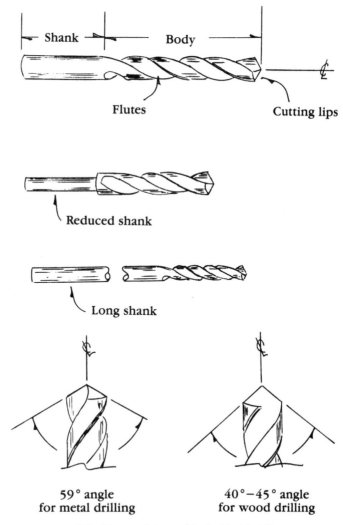

2-3 Nomenclature of typical twist drills.

Table 2-1   Three Systems Denoting Twist Drill Sizes

NUMBER				FRACTION		LETTER	
#	Decimal	#	Decimal	Fraction	Decimal	Letter	Decimal
80	.0135	39	.0995	1/64	.0156	A	.234
79	.0145	38	.1015	1/32	.03125	B	.238
78	.016	37	.104	3/64	.046875	C	.242
77	.018	36	.1055	1/16	.0625	D	.246
76	.02	35	.11	5/64	.078125	E	.25
75	.021	34	.111	3/32	.09375	F	.257
74	.0225	33	.113	7/64	.109375	G	.261
73	.024	32	.116	1/8	.125	H	.266
72	.025	31	.12	9/64	.140625	I	.272
71	.026	30	.1285	5/32	.15625	J	.277
70	.028	29	.136	11/64	.171875	K	.281
69	.0292	28	.1405	3/16	.1875	L	.29
68	.031	27	.144	13/64	.203125	M	.295
67	.032	26	.147	7/32	.21875	N	.302
66	.033	25	.1495	15/64	.234375	O	.316
65	.035	24	.152	1/4	.25	P	.323
64	.036	23	.154	17/64	.265625	Q	.332
63	.037	22	.157	9/32	.28125	R	.339
62	.038	21	.159	19/64	.296875	S	.348
61	.039	20	.161	5/16	.3125	T	.358
60	.04	19	.166	21/64	.328125	U	.368
59	.041	18	.1695	11/32	.34375	V	.377
58	.042	17	.173	23/64	.359375	W	.386
57	.043	16	.177	3/8	.375	X	.397
56	.0465	15	.18	25/64	.390625	Y	.404
55	.052	14	.182	13/32	.40625	Z	.413
54	.055	13	.185	27/64	.421875		
53	.0595	12	.189	7/16	.4375		
52	.0635	11	.191	29/64	.453125		
51	.067	10	.1935	15/32	.46875		
50	.07	9	.196	31/64	.484375		
49	.073	8	.199	1/2	.5		
48	.076	7	.201				
47	.0785	6	.204				
46	.081	5	.2055				
45	.082	4	.209				
44	.086	3	.213				
43	.089	2	.221				
42	.0935	1	.228				
41	.096						
40	.098						

for metal (FIG. 2-3). Some operators regrind cutting edges, but it's possible to buy units that are ready for wood drilling.

One of the reasons for the popularity of twist drills is their extensive range in sizes. Often, one is used simply because another type of drilling tool is not available for the hole size that is needed.

Three different systems are used to denote the sizes of the twist drills: number, fraction, and letter (TABLE 2-1). Smallest diameters are listed in wire gauge sizes running from smallest, Number 80 (.0135) to largest, Number 1 (.228). Letter sizes cover from A (.234) to Z (.413). The third set, which states diameters in fractions, starts at 1/64 inch and increases by that dimension up to 1/2 inch; it overlaps its two kin but does not duplicate. TABLE 2-2 lists more decimal equivalents of fractions that can be helpful when determining a drill size, and for other mathematical problems.

There are twist drills designed for particular applications. The example shown in FIG. 2-4 is a recent innovation and does a fine job forming holes for screws. There are other cutters that make preparing for screws a fast and accurate procedure, but they are not true drill bits. (These will be demonstrated in Chapter 5.)

Twist drills may be purchased individually or in sets. The latter is the practical approach because overall cost will be less and an assortment will allow you to be

Table 2-2 Decimal Equivalents of Fractions

Fraction	Equivalent	Fraction	Equivalent	Fraction	Equivalent
1/64	.015625	1/32	.03125	3/64	.046875
1/16	.0625	5/64	.078125	3/32	.09375
7/64	.109375	1/8	.125	9/64	.140625
5/32	.15625	11/64	.171875	3/16	.1875
13/64	.203125	7/32	.21875	15/64	.234375
1/4	.250	17/64	.265625	9/32	.28125
19/64	.296875	5/16	.3125	21/64	.328125
11/32	.34375	23/64	.359375	3/8	.375
25/64	.390625	13/32	.40625	27/64	.421875
7/16	.4375	29/64	.453125	15/32	.46875
31/64	.484375	1/2	.5	33/64	.515625
17/32	.53125	35/64	.546875	9/16	.5625
37/64	.578125	19/32	.59375	39/64	.609375
5/8	.625	41/64	.640625	21/32	.65625
43/64	.671875	11/16	.6875	45/64	.703125
23/32	.71875	47/64	.734375	3/4	.75
49/64	.765625	25/32	.78125	51/64	.796875
13/16	.8125	53/64	.828125	27/32	.84375
55/64	.859375	7/8	.875	57/64	.890625
29/32	.90625	59/64	.921875	15/16	.9375
61/64	.953125	31/32	.96875	63/64	.984375

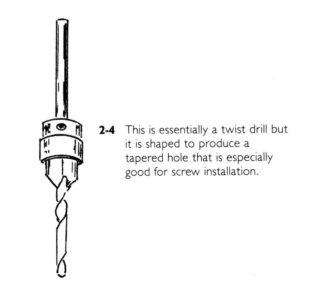

2-4 This is essentially a twist drill but it is shaped to produce a tapered hole that is especially good for screw installation.

prepared for various hole sizes. A set of fractional size drills is a good beginning for a home shop. Buy them with a drill stand or a storage case. The units will be stamped with a size at each hole so a particular one can be easily selected.

BRAD POINT BITS

Brad point bits, often called machine spur bits, are precision cutters that produce optimum results in wood, whether working parallel to or across the grain (FIG. 2-5). The point on the bit makes it easy to locate at the center of the hole and does

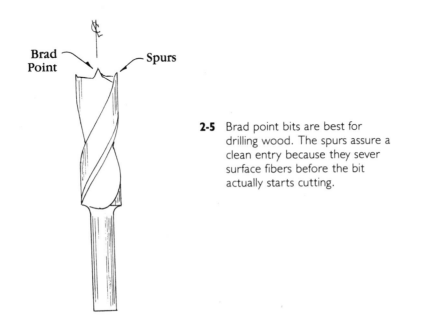

Brad Point   Spurs

2-5 Brad point bits are best for drilling wood. The spurs assure a clean entry because they sever surface fibers before the bit actually starts cutting.

a good job of keeping the cutter from wandering. The spurs make initial contact, cutting clean through surface fibers, *before* the bit actually begins removing wood. This results in a hole that is exact in size and that leaves clean edges both entering and leaving the work. Few cutters do as good a job as brad point bits for chores like drilling for dowels.

Common sets start at $1/4$ inch but $1/8$-inch and $3/16$-inch sizes can be found. Make or buy a partitioned container to hold more than the number of bits you buy initially (FIG. 2-6).

SPADE BITS

Spade bits (FIG. 2-7) are good wood-drilling tools, especially when extra-large holes are required. They have long, sharp points for initial penetration and slim shanks that provide plenty of room for waste material. They have flat blades and, unlike other bits, they essentially do more scraping than cutting. Quality products have blades with relieved edges that create cleaner, cooler cutting.

Spade bits operate most efficiently at higher-than-average speeds. Even the largest one, usually $1^1/2$ inch, should turn between 1500 and 2000 RPM. These

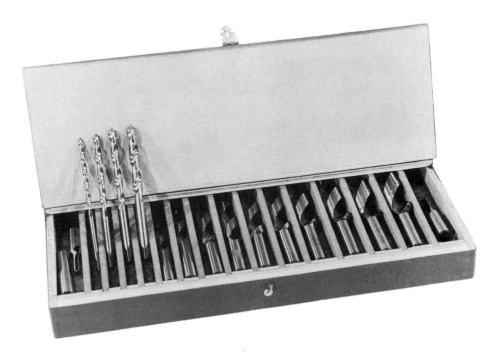

2-6 A storage case that can be made for brad point bits.

2-7 Good spade bits have sharp points and relieved edges. Flats on the shank are a good idea because they will prevent the bit from slipping in the chuck.

bits don't enter the work as cleanly as, for example, brad point bits, and the work-piece must be firmly set on a backup when drilling through to avoid excessive feathering.

The bit's concentricity is critical, but this is easy to check when you shop. Place the shank on a counter and roll it under the palm of your hand. If the point on the bit describes a small circle as you roll it back and forth, concentricity is not what it should be. With this same test you will be able to tell if the shank of the bit is or isn't true.

EXPANSIVE BITS

Expansive bits (FIG. 2-8) are widely available for use in a hand brace, but can be used in a drill press if the tool has a brad point. The advantage of such a bit is the

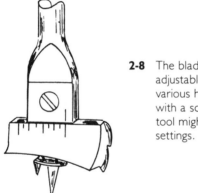

2-8 The blade on an expansive bit is adjustable so it can be used for various hole sizes. Always check with a scale even though the tool might be calibrated for settings.

single cutter that can be moved in and out to form any size hole within its capacity. Thus, an odd hole size is possible because the settings between minimum and maximum are infinite.

The tools are generally available in two sizes, each with two cutting blades. Capacities are: $5/8$ inch to $13/4$ inch and $7/8$ inch to 3 inch. Because the cutters pose an off-center load, it's essential to use them at a slow speed and with the work securely clamped.

EXTENSION BITS

Extra-deep holes, like those you might need for a lamp base, can be accomplished with extension bits like those shown in FIG. 2-9. Quill extension, of course, is also a factor in hole depth, but there is a technique to use to lengthen the hole. First drill a maximum-depth hole, then retract the bit. Raise the table so that the bit will bottom in the hole already drilled in the work, then drill again. This can be repeated to the full length of the bit.

FORSTNER BITS

Forstner bits (FIGS. 2-10 and 2-11) are not newcomers, but they seem to have been suddenly discovered by home shop drill press users—and with good reason. They produce true, flat-bottom, blind holes—whether cutting across or with the grain—and will bore cleanly through thin stock and even veneers.

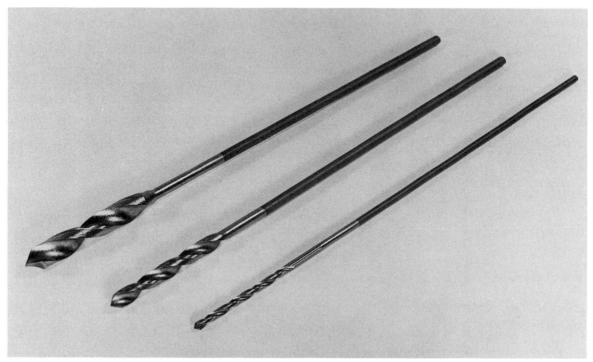

2-9 Extension bits are used to drill very deep holes.

2-10   Forstner bits do impressive work. Even large ones can be used efficiently in an average drill press.

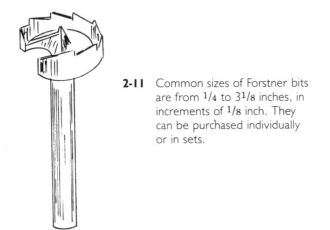

2-11   Common sizes of Forstner bits are from $1/4$ to $3^1/8$ inches, in increments of $1/8$ inch. They can be purchased individually or in sets.

A unique characteristic of the bits is that they don't depend on the central brad point for alignment; they can actually be guided by peripheral cutting edges. Thus you can bore at the edge of a workpiece with the brad point outboard to form an arc of a circle, oval shapes, and curved openings. Even difficult pocket holes, needed to attach table tops to rails, can be accomplished without fear that the bit might wander (FIG. 2-12).

2-12 Forstner bits have unique drilling characteristics. They cut cleanly even when the center point is not located on the stock.

Forstner bits are available in a range of sizes. Even the largest—3 or $3^1/8$ inches—cut efficiently in the average drill press. Anyone interested in making battery-powered clocks can buy a Forstner bit that will continually produce precise holes for the clock units.

HOLE SAWS

Hole saws are usually relied on for very large holes. The type available in units of fixed diameter look much like steel cups with saw-teeth rims (FIG. 2-13). These may be purchased individually or in sets that consist of an assortment of saws and an interchangeable mandrel. Depth of cut is a consideration. Some products are limited to $3/4$-inch stock while others can penetrate $1^1/2$-inch material. A bonus feature is that the tools form a hole by removing a disc—circular pieces that can be used, for example, as wheels for toys. Because the mandrel operates with a pilot bit, the wheels will already have an axle hole.

The metal for hole saws is usually tempered so the saws can be used on metal as well as wood. Some have teeth of tungsten carbide and so are better suited for metal working. Hole saws work best at slow speeds; the larger the tool, the slower the speed. It's good practice to start at the slowest speed, regardless of the tool's size. Then increase speed until the saw is cutting consistently without overheating and without screeching, the latter being a warning sign that you are not operating correctly.

Use this technique when you require a hole through stock that is thicker than the saw's capacity. First, use a bit that equals the diameter of the tool's pilot to drill through the work. Next, cut into one surface to a depth slightly more than half of the stock's thickness. Finally, invert the stock so you can cut into the opposite surface. The predrilled hole will guide the saw so the cuts will meet in good style.

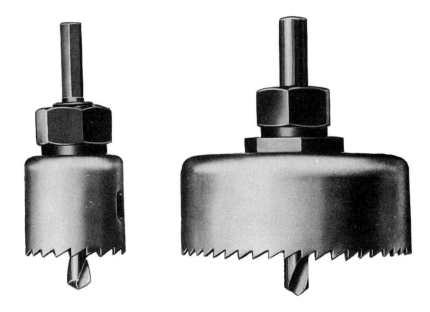

2-13   Fixed-diameter hole saws resemble steel cups with serrated rims. A set of four or five saws will be supplied with a single, interchangeable mandrel. The pilot bit is replaceable.

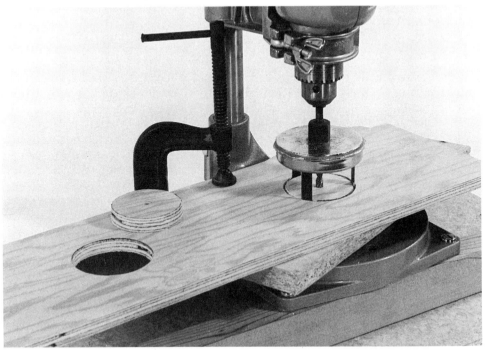

2-14   This type of hole saw works with a trio of blades that are adjustable for various diameters. The tool often comes with two sets of blades—one for wood, the other specially tempered for metal.

A type of adjustable hole saw is shown in FIG. 2-14. Individual blades are moved in concert toward or away from the midpoint by a perimeter, lockable control ring. An advantage is that the one tool can form any size hole. A disadvantage is that accuracy will depend on how carefully the tool is adjusted.

Another type of adjustable hole saw is displayed in FIG. 2-15. In this case, the tool is supplied with saw-edged bands of various diameters that lock in circular grooves in the head. The unit is equipped with a spring-loaded ejection system that helps to remove captured discs.

2-15 An assortment of circular, band-type blades make this hole saw design flexible in terms of hole sizes. The blades are locked into grooves that are in the head. The product is available with short or long blades.

FLY CUTTERS

Fly cutters can form small holes but they are especially useful when very large holes or discs are needed. The units rotate a vertical bit that is at the end of a horizontal arm (FIG. 2-16). Because the arm is adjustable, the bit can be set to cut any circle within its capacity.

A recent innovation is shown in FIG. 2-17. This design employs two bits to make cutting easier. This makes sense, but only if you organize the tool so both cutters are following the same path. The horizontal bar is calibrated for bit settings but shouldn't be trusted blindly. Use a tape for checking before cutting. Having two cutting bits prompts some interesting applications, which will be discussed further in Chapters 5 and 13.

Figure 2-18 shows a fly cutter with a different operating style. The unit turns a slanted cutting blade so that the hole is formed without leaving a central disc. The action is seen more easily in cross section (FIG. 2-19). Whether this is an advantage over outboard bit fly cutters is debatable, but it is good for making overlapping cuts for decorative work.

2-16   A typical fly cutter used to form sheet metal discs. (Note that the pilot bit has been removed so the discs will not have a center hole; this is not the normal procedure.)

PLUG CUTTERS

Plug cutters are unique tools that form short or long cylinders. These can be used as dowels in joints or as plugs that are set in counterbored holes to conceal screws (FIG. 2-20.). Whether you go one way or the other depends on the style of cutter

2-17 This fly cutter design employs twin blades. Calibrations on the arm help to position the cutters.

you use. Some are limited in depth of cut and are usable mostly to form screw concealment plugs (FIG. 2-21). Others form longer cylinders and are fine for making dowels that can be used in joints.

An advantage of plug cutters is that you can use them on wood that matches the project material and, in the case of screw concealment, you can cut so the grain is compatible with the flow of adjacent surfaces. This is difficult to do with store-bought examples, assuming that you can even find them in the wood species you need. Chapter 7 discusses applications for plug cutters in more detail.

COUNTERSINKS

Countersinking is required for flat-head screws that must be driven flush with the work's surface. The countersink itself (FIG. 2-22) is a cutting tool that operates something like a drill bit, but forms an inverted cone of limited depth for the screw head. Countersinking is done to full screw head depth on hardwood and slightly less on softwood, because softwood will allow the screw to seat itself flush when you are installing it.

It's acceptable to judge cut depth by eye when making just one countersink. When you need many similar cuts, which is usually the case, it is best to establish a depth control by utilizing the drill press stop rod feature.

2-18 This fly cutter works with a slanted blade so the operation does not produce a disc.

2-19 Cross-section view shows how the slanted blade fly cutter works.

2-20 This type of cutter can be used to form short plugs or cylinders up to 2 inches long. The advantage of plug cutters is that they can be used to cut with or across the grain, and in any wood species.

1/2-inch long
diameters of: 3/8, 7/16, 1/2, 5/8 inches

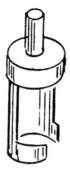

3/4-inch long
diameters of: 3/8, 1/2, 5/8 inches

3¹/8-inch long
diameters of: 3/8 to 1 inch

2-inch long
diameters of: 1/4, 3/8, 1/2 inches

2-21 Types of plug cutters that are available and the sizes of cylinders they can produce.

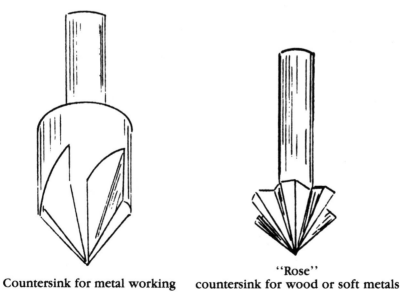

Countersink for metal working   "Rose"
countersink for wood or soft metals

2-22 Two common types of countersinks.

Chapter **3**

Major accessories

Proper accessories add to drill press applications and can make many jobs easier. Often an accessory provides the means to do procedures that ordinarily require an additional, often expensive, machine. You can manufacture many accessories in your own shop. These will be described and detailed in many of the following chapters. Some of them are just not available from retailers even if you want to buy them; others serve in the same fashion as store-bought units but at considerably less expense.

INTERCHANGEABLE SPINDLES

If your machine is designed for interchangeable spindles, consider adding an extra one or two for better handling of specific applications. Supplementary spindles will save wear and tear on the supplied unit and on the drill press as a whole. Often, a special spindle makes it possible to keep cutter and chuck closer to bearings than you could otherwise.

Two common spindles are shown in FIG. 3-1. The one with the geared chuck will match the capacity of the tool holder that comes with the machine and is used for general drill press work. The other has a special, integral chuck that grips tools with 1/2-inch shanks by means of set screws. This design provides an efficient method of gripping cutters that develop side thrust and that should not be held in a conventional three-jaw chuck—among them, router bits. Other tools the spindle can be used for include drum sanders, fly cutters, and rotary planers.

It is not, of course, a necessity to buy extra spindles. Most manufacturers offer adapters that facilitate non-drilling applications and that are secured to the conventional spindle. These include router chucks, shaper adapters, extra three-jaw chucks, and special holders for drill bits that have tapered shanks (FIG. 3-2).

3-1 Common extra spindles that can be used in place of the standard one.

3-2 Adapters that attach to the spindle in place of the conventional chuck are available for operations like routing and shaping.

FOR MORTISING

The combination of mortise and tenon provides a strong and durable woodworking joint (FIG. 3-3). The tenon, which is a square or rectangular projection, can be made on a table saw. The mortise, a matching cavity that is the more difficult of the components to produce, is an easy and common drill press function.

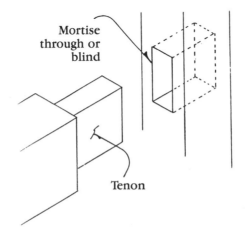

Mortise
through or
blind

Tenon

3-3 The mortise-and-tenon joint. The tenon is a table saw function; the mortise is done with special tools on a drill press.

The secret lies in special cutters that are called mortising bits and chisels. The bit forms a hole much like a drilling tool, but it is encased and rotates within what is really a four-sided chisel. The bit removes the bulk of the waste while the chisel cleans out the corners. The result is a square hole (FIG. 3-4). The hole can be used as-is for small tenons or can be elongated by making overlapping cuts.

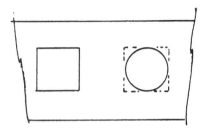

3-4 Mortising bits and chisels work together to form square cavities: The bit removes waste and the chisel cleans out corners. The shape can be made wider or longer with additional overlapping cuts.

In order to use the chisels and bits, it's necessary to have a special casting that is secured in some fashion to the end of the quill. The component is always part of a mortising kit that includes a special fence and hold-down similar to the example shown in FIG. 3-5. An exception is the kit available for the multipurpose Shopsmith. In this case, the machine's rip fence is utilized so the extras required are just the casting and hold-down (FIG. 3-6).

3-5 Typical mortising attachment consists of a casting that grips the chisel, plus a fence and hold-downs. The bit, which rotates inside the chisel, is gripped by the standard chuck.

3-6 The mortising setup on a Shopsmith utilizies the fence that is standard equipment for the machine in all its modes.

Mortising bits and chisels are sold as teams; the chisel to match the bit. Common sizes include $1/4$ inch, $3/8$ inch, and $1/2$ inch. The technique of mortising will be covered in Chapter 8.

FOR SHAPING

Shaping is a woodworking technique used to form decorative edges on components such as tabletops. It can also be used, to some extent, to form joints.

Shaping is a practical drill press application as long as you are aware of two factors: The machine's top speed doesn't match the RPM of an individual shaper, and the cutter-holder adapter that attaches to the spindle is above the table instead of being under it, as on a true shaper. However, reduced speed can be compensated for by slowing the feed rate (how fast you move the work), and because having the spindle above the table isn't critical for many shaping operations, you can accomplish creditable work.

Shaping requires an accessory kit that includes a special table to accommodate twin fences that can be moved independently forward and back. Figure 3-7 shows a typical setup. Because this is a Shopsmith model, the fences attach directly to the table that is basic to the machine in all its modes.

3-7 Shaping arrangement on a multipurpose machine. The accessory consists of individually adjustable fences that are secured to the standard table.

Shaping tables and fences are not available for all drill presses, but you can make one similar to the unit shown in FIG. 3-8. Like professional models, it has individually adjustable fences whose bearing surfaces can be moved laterally to minimize the opening around the cutter.

Cutting is done with three-lip shaper cutters (FIG. 3-9). These are just a few of the shapes and sizes available. In addition to the cutters, an adapter is needed so they can be safely mounted on the spindle in place of the standard chuck. What

3-8 A shaping jig that you can make. The unit consists of a special table and fences that can be moved laterally and to and fro.

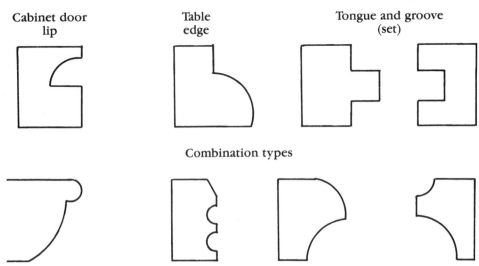

3-9 Shaper cutters for drill press application are of the three-lip design. These examples are just a few of the many profiles that are available.

3-10 Drill press adapters. The one on the right is for three-lip shaper cutters. A special chuck for router bits is at the left. Adapters are peculiar to the drill press they must be used on, so they will all look different.

the adapter looks like will depend on the machine's requirements. It will be, essentially, a threaded shaft that is grooved to receive a keyed washer and a lock nut. Collars that are used over, under, or between cutters—depending on the job—might be supplied with the adapter or they will be available as accessories. A typical adapter, together with one that is used for routing, is shown in FIG. 3-10.

The techniques of shaping and making accessories will be covered in Chapter 10.

FOR ROUTING

Drill press routing is done with the same bits used in a portable router. The work can be done freehand as shown in FIG. 3-11, or by using guides for straight-line work. There are no table accessories for routing operations, but this is a problem that is easily solved. Straight cuts can be accomplished by clamping a straight piece of wood to the drill press table as a fence, or by making a jig like the one shown in FIG. 3-12 for general use.

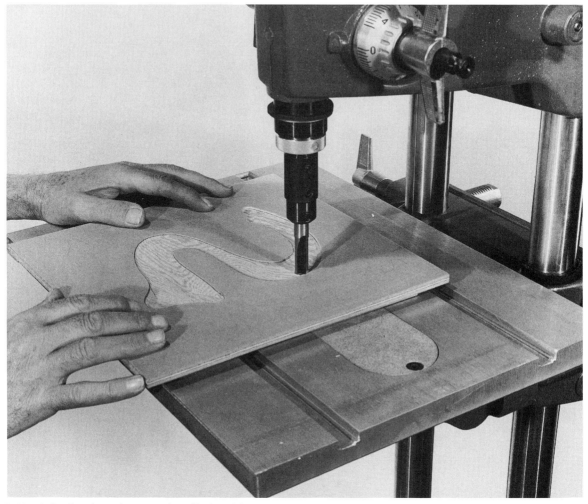

3-11 A freehand routing operation. Bits used for routing are the same ones that are used in a portable router.

Router bits must be held in the special chuck that was shown together with a shaper adapter in FIG. 3-10. Routing serves many purposes and will be explained further in Chapter 9.

FOR DRUM SANDING

Drill press drum sanding can smooth the way to a lot of clean edges, whether straight or curved. The major accessories required are the sanding tools themselves, which are available in various shapes and sizes (FIG. 3-13). In addition, the machine's table should provide for an access hole so that by moving the quill up and down, the entire sanding area of the drum can be utilized.

This is accomplished on a Shopsmith by substituting a special table insert for the regular one (FIG. 3-14). Most drill press tables, however, are not designed for

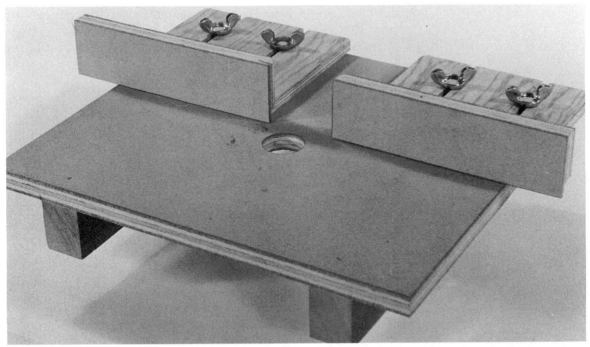

3-12 Straight routing calls for a guide so the work can be moved accurately. The guide can be a simple, clamped-on fence or a more sophisticated, more flexible unit like this homemade one.

3-13 Sanding drums are available in many shapes and sizes. Most of them work with pre-formed abrasive tubes; others require attaching strips of abrasive material.

3-14 Drum sanding should be organized so the drum can be moved up and down. A special insert with an access hole for the sander permits this on a multipurpose tool.

the application. Many operators will simply swing the table just far enough aside to clear the drum so they still have a work supporting platform. A better solution is to provide a raised platform, with a center hole for the drum that can be clamped to the standard table (FIG. 3-15).

A variety of drum sanding operations and how to make accessories for the applications will be covered in Chapter 11.

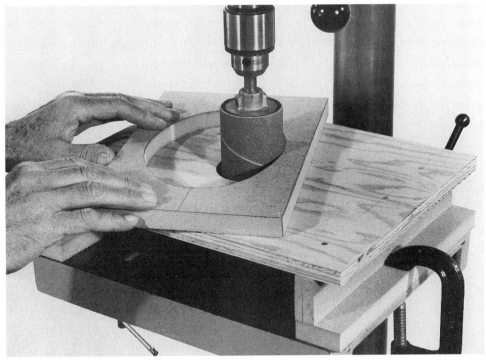

3-15 A simple elevated table that can be clamped in place works very well on a conventional drill press. Being able to move the drum vertically allows full utilization of the abrasive surface.

Chapter **4**

Safety

Safety must be a top priority in any workshop. Damage often results when the operator does not pay attention when using tools or disobeys good practice rules, or simply develops overconfidence along with expertise. The degree of safety does not necessarily increase in proportion to the time spent using the machine.

Many craftsmen forget safety rules in their eagerness to get started. This is breaking the paramount safety rule! There is little point in rushing toward proficiency at the risk of lacerated fingers.

It's important to realize that the drill press is not a "safe" tool. There are potential hazards in all shop work, regardless of which machine is used. Comparisons are often made in areas of degree of damage that is possible with various tools. But the goal is to avoid damage entirely.

Because different brands of tools have particular characteristics, the first step is to thoroughly understand the one you are using. This is simply a matter of studying the owner's manual and becoming familiar with what the tool can and can't do. It's possible that you might be advised against non-drilling chores like routing and shaping and to avoid the use of such tools as fly cutters. You must obey whatever limitations are imposed.

Drill presses are not equipped with guards so you must know before you start the best place for your hands and how the cutting tool operates. In all situations, whatever you can do or make to create a wall between you and a danger zone makes sense.

The homemade safety shield mounted on a drill press in FIG. 4-1 is an example. It's a very practical guard that should be used, for example, when doing routing and shaping operations. The plastic shield mounts on a U-shaped bracket that is attached with the same bolt and nut that tightens the split casting that supports the depth stop rod. Construction details of the unit are shown in FIG. 4-2. The prototype was made for a Delta product so be sure to check the dimensions of the

4-1 The drill press is not equipped with a guard, but you can make one. This guard has been set up for a routing chore.

brackets against your own tool before cutting parts. Figure 4-3 offers an alternate mounting arrangement if you find that the U-shaped design is not feasible.

SECURING THE WORK

Because the drill press operates by rotating cutters, there is a tendency for work-pieces to spin as well. The action is more pronounced on hard materials like metal, and when oversize holes are being formed. However, the possibility of movement always exists and should be guarded against. When the workpiece is long, it can be situated so its rear edge abuts the left side of the column, preventing the tendency to twist with the cutter. A more general rule is to secure the work to the table with clamps (FIG. 4-4).

Various types of work-holding devices are available for the drill press. Figure 4-5 shows a clamping design that locks to an edge of the table and is flexible enough so whatever it grips can be located under the spindle.

Drill press vises (FIG. 4-6) are common accessories, and while they are offered primarily for metal working, there is no reason why they can't grip wood. The

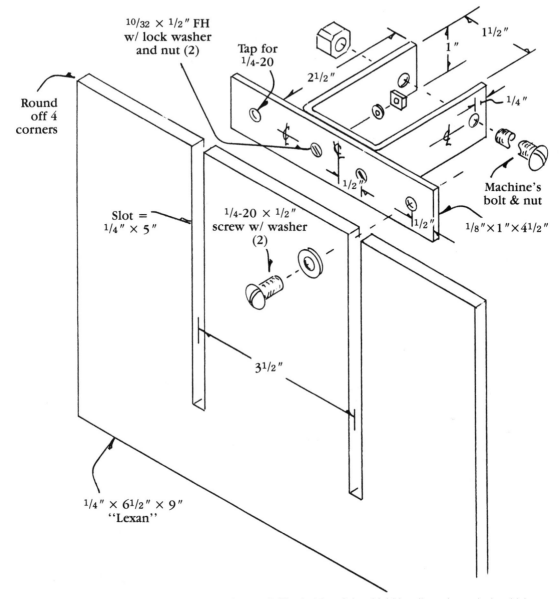

Round off 4 corners

$10/32 \times 1/2''$ FH w/ lock washer and nut (2)

Tap for $1/4$-20

$2^{1}/2''$

$1''$

$1^{1}/2''$

$1/4''$

$1/2''$

Machine's bolt & nut

$1/2''$

$1/2''$

$1/8'' \times 1'' \times 4^{1}/2''$

Slot = $1/4'' \times 5''$

$1/4$-$20 \times 1/2''$ screw w/ washer (2)

$3^{1}/2''$

$1/4'' \times 6^{1}/2'' \times 9''$ "Lexan"

4-2 Construction details for the homemade guard. The height of the shield is adjusted to suit the thickness of the workpiece.

products are often made with projecting flanges so clamping them in position is feasible (FIG. 4-7).

Don't attempt to hand-hold small pieces. When necessary, shape a scrap piece of wood so the component can be secured to it and the improvised holder can be clamped to the table (FIG. 4-8). A woodworking clamp, called a *handscrew*, was modified by forming matching Vs in its jaws. This is a good way to grip short

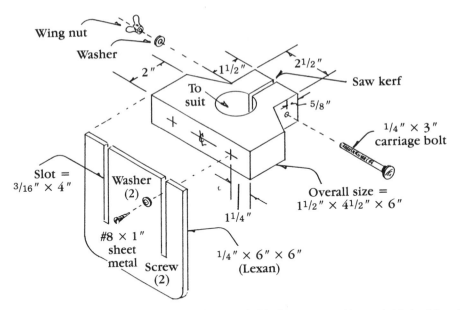

Wing nut

Washer

2"

1¹/₂"

2¹/₂"

To suit

Saw kerf

5/8"

¹/₄" × 3"
carriage bolt

Slot =
3/16" × 4"

Washer
(2)

#8 × 1"
sheet
metal   Screw
(2)

1¹/₄"

¹/₄" × 6" × 6"
(Lexan)

Overall size =
1¹/₂" × 4¹/₂" × 6"

4-3 If the U-shaped attachment bracket is not suitable for your machine, substitute this split-clamp design that secures to the end of the quill.

4-4 The workpiece must always be clamped securely.

4-5 One type of clamping accessory. The tool may be secured at any side of the table and is flexible enough so that situating the workpiece under the spindle poses no problem.

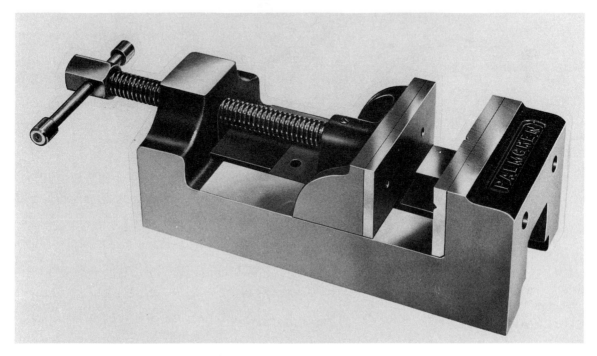

4-6 A typical drill press vise.

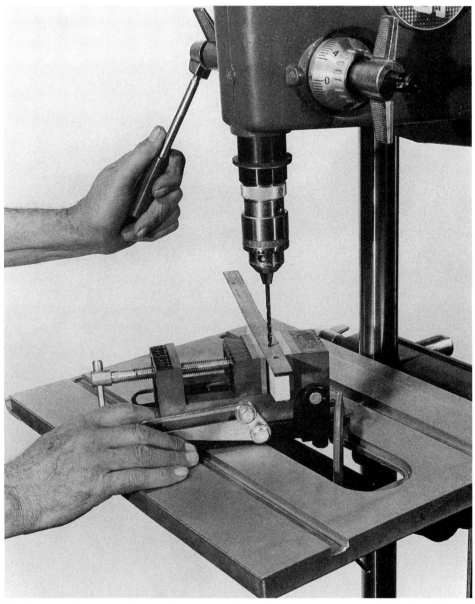

4-7 Some drill press vises have extending base flanges so they may be clamped in place. This unit is designed so the jaw component can be tilted. Thus, it can also be used for angular drilling.

squares or cylinders for concentric drilling. The handscrew can be clamped to the drill press table.

CHUCK AND KEY

The common drill press chuck is a three-jaw design. Turning its barrel causes its jaws to move uniformly against the shank of the cutter. You can turn the barrel by

4-8 A clamping device to hold squares or cylinders for concentric drilling, made by sawing matching Vs in the jaws.

hand to make initial contact, but the final tightening must be done with the chuck key. If the cutter is not secure it will simply spin in the chuck, causing abnormal wear on cutter and chuck and present the danger that the tool might fly out and harm the work or you. The jaws of the chuck should bear against the flat part of any tool shank that is so designed.

Always remove the key. There have been incidents where keys inadvertently left in the chuck become dangerous projectiles when the machine is turned on.

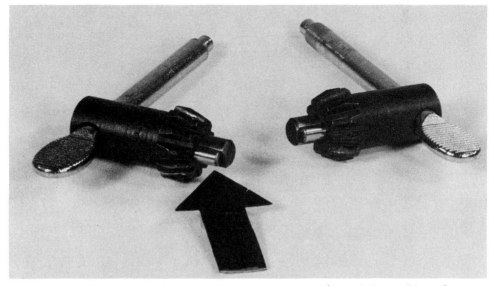

4-9 Some chuck keys incorporate a spring-loaded device so it's nearly impossible to forget to remove them from the chuck.

Some keys are designed to help prevent this danger. The example in FIG. 4-9 incorporates a spring-loaded shaft that will disengage the key when you relax pressure. Some drill presses are designed so they can't be turned on unless the key is inserted in some fashion to serve as a preliminary switch.

PROTECTING EYES, EARS, AND LUNGS

Always wear adequate safety goggles or a face mask to protect vision. Workers realize the importance of protecting eyes, but many are lax about protecting ears and lungs. Headphone-type protectors for hearing are as important as any safety device. High frequencies can be generated by high-speed electric motors and by air movements and work sounds that are part of many drill press operations. Routing and shaping are prime examples. A single exposure might not do harm, but effects during frequent and prolonged usage of a machine are cumulative and contribute to potential hearing damage. Quality hearing protectors will shield you from damaging high frequencies but will not eliminate the normal woodworking noises that you should hear.

Anytime you work on wood, and many nonwood materials, you create dust and particles that belong in a vacuum cleaner, not your lungs. Wear a dust mask,

4-10   Dust masks should be worn to protect nose and lungs from airborne particles. This disposable type is easy and comfortable to wear, but it is strictly a DUST mask.

even when you feel the operation is causing minimum waste. Masks shown in FIG. 4-10 are disposable; easy and comfortable to wear, even in conjunction with safety goggles. Be aware that products like this do not protect against toxic fumes. Dust masks are available that use a material to filter the air. The filters should be replaced or cleaned as often as necessary.

PROPER SHOP ATTIRE

It is important to have special shop clothes, for cleanliness as well as safety. Wear pants and shirts that fit snugly. Billowing materials can snag against tool parts and can even be caught by rotating cutters. Heavy, nonslip shoes, preferably with steel toes, are advisable. Gloves are not practical and anything like a necktie or scarf is a potential hazard. Rings, wristwatches, bracelets and similar items must not be worn when doing any kind of shop work. Provide a cover for your hair, whether long or short, to protect against dust as well as for safety.

SHOP ENVIRONMENT

Maintain the shop as you would a living area. Tables, benches, the floor, and tool surfaces should always be in pristine condition. Accept a shop type vacuum cleaner as an important piece of equipment. The product is available in various sizes and price ranges. Most models will have a port for the hose so the unit can function as a blower to remove dust and dirt from tight corners.

Frequently remove the dirt and gummy substances that accumulate on tool tables. Cleaning solvents, used carefully according to the directions on the container, are often sufficient to return the table to proper condition. In extreme cases, you can go over the table with a pad sander that holds a very fine emery abrasive. The weight of the sander is enough to supply necessary pressure, so don't force as you move it about. Wipe the table with a lint-free cloth and then apply a coat of paste wax. Rub the wax to a fine polish after it is dry. Wax frequently to keep tables protected and to help workpieces move smoothly and easily. The waxing suggestion applies to tables of jigs you make as well as to regular tool tables.

SHOP BEHAVIOR

The primary rule of shop behavior is: Always be alert. Give complete attention to your work. Don't work in the shop if you are tired or upset, or after taking medicine or an alcoholic drink.

Don't work with dull tools. Results will be poor. Also, you will have to exert extra pressure to keep the work moving or the tool cutting, which creates the possibility that your hands might slip.

Overreaching, no matter what the operation or the tool, is bad practice. You can be thrown off balance or place your body in a hazardous position. Have someone help you when you are working on a piece that is too large for you to control. Just be sure to explain the procedure to the helper so he or she will know how to cooperate. An outboard support stand is a good accessory to make for use when working solo (FIG. 4-11 and the materials list on p. 63).

Have the tool plugged in only when you are using it. If you are cleaning the

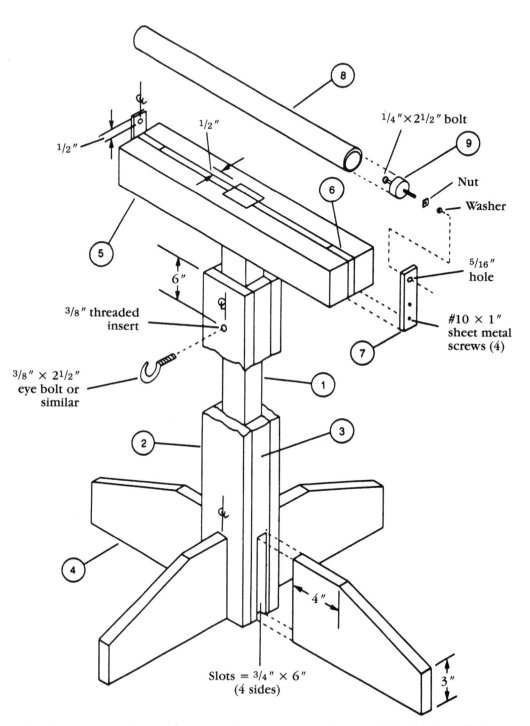

1/2"

1/2"

1/2"

8

1/4" × 2 1/2" bolt

9

Nut

Washer

6

5/16" hole

5

6"

3/8" threaded insert

#10 × 1" sheet metal screws (4)

3/8" × 2 1/2" eye bolt or similar

7

1

2

3

4

4"

Slots = 3/4" × 6"
(4 sides)

3"

4-11 A roller-top, outboard support stand that you can make. Because it is height-adjustable, it can be used with a variety of shop tools.

4-11 Continued.
Materials List

Key	Part	No. of Pieces	Size (in inches)			Material
1	Post	1	$1^1/2 \times 2$		$\times 30$	Hardwood
2	Case	2	$3/4 \times 3^1/2$		$\times 30$	Plywood
3	Case	2	$3/4 \times 1^1/2$		$\times 30$	Plywood
4	Feet	4	$3/4 \times 6$		$\times 11$	Plywood
5	Roller support	2	$1^1/2 \times 2^1/2$		$\times 18^1/2$	Hardwood
6	Filler	2	$1/2 \times 2$		$\times 2^1/2$	Hardwood
7	Holder	2	$1/8 \times 1$		$\times 4^1/4$	Aluminum
8	Roller	1	$1^1/2$	O.D.	$\times 18$	Rigid tubing
9	Plug	2	$1^1/2 \times 1^1/2$	$\times$	$1^1/2$	Hardwood

machine or making a setup that takes time, be cautious enough to remove the plug from the power source. Check to be sure the switch is in the off position before you plug in. Don't leave the tool running while you attend to another chore, regardless of how little time is involved. Don't leave cutters mounted in the chuck when you are not working. Always wait for the cutting tool to stop before you move away from the machine. If the tool has a safety key, as some do, take it with you when you stop working, or hide it. Preventing unauthorized use is an important safety factor.

The workshop is not the place for socializing. Don't combine working and visiting. Let friends and neighbors know that they must not barge into your shop if they hear a tool running. You don't want to be startled.

TOOL PRACTICE

Maintain the tool in pristine condition and be sure to follow the manufacturer's instructions if lubrication is necessary. When buying commercial accessories, be sure they can be used safely on the tool you own. Don't disregard any advice from the manufacturer concerning the machine's limitations.

If the tool makes strange noises or stops operating as it should, check the owner's manual for a possible solution to the problem. If you can't find one, contact the supplier to determine what steps to take.

When making jigs or accessories, check the supplied dimensions against the tool you will use them on. Homemade units are important aids. Make them carefully and treat them with respect.

Don't be too generous in allowing others to use your shop and equipment. Remember that you are liable for their safety.

ELECTRICAL CONSIDERATIONS

Full-size drill presses are usually supplied with a three-conductor cord and a grounding plug that should be used in a matching three-conductor grounded out-

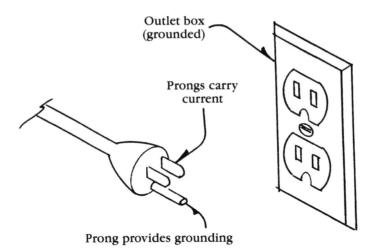

4-12 A three-prong plug is used in a matching, grounded receptable. *Never* remove or alter the grounding prong in any way.

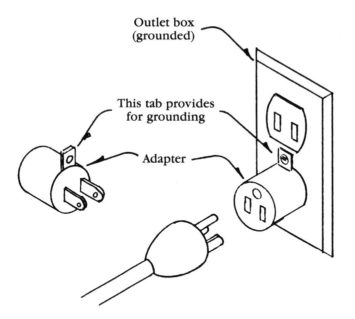

4-13 An adapter that will receive a three-prong plug can be attached like this to a two-hole outlet. This temporary measure will provide grounding only if the wiring to the outlet provides grounding.

let (FIG. 4-12). If the outlet that is available for the machine is the two-prong type, an adapter can be used between the plug and the outlet (FIG. 4-13). *Never* remove or alter the grounding prong in any way.

There are several important factors to be aware of regarding adapters. Consider them a temporary solution; use them only if you are sure that the two-prong receptable is correctly grounded. If it isn't the tool won't be either. If there is doubt about whether the system is correctly wired, have it checked by a qualified electrician. Incidentally, the use of the adapter is not permitted in Canada.

Be sure to check the manufacturer's information concerning other electrical factors: sizes and types of fuses or circuit breakers, types and sizes of extension cords to use should you ever use the tool away from its standard outlet, and so on. Like other power tools, the drill should not be used in a damp or wet environment.

Chapter **5**

Basic drilling

Forming through or blind holes is the primary function of a drill press. The procedure is simple: Secure a bit in the chuck, place the workpiece, and pull down on the feed lever to extend the quill. Minimizing disruption of surface fibers where the bit enters and leaves the work, and achieving a hole that has smooth walls and that is accurately located are quality factors that depend on the operator's full attention.

A dislocated hole can ruin a component and one that is ragged or slightly over or undersize won't receive a dowel as it should. A hole that is not square to the work's surface will cause considerable frustration when the part it must receive is placed or when—in the case of a dowel joint—the mating parts are placed together. Even a screw hole that is slightly slanted can be a nuisance. These and other negative factors are easily avoided simply by working carefully.

ALIGNMENT

The one important alignment factor for normal drilling is to be sure the angle between the table and the bit is 90 degrees. A good way to check is to secure a length of steel rod in the chuck and then use a square to determine if the angle is correct (FIG. 5-1). Do this with the indexing pin (if the machine has one) in place. Then, if necessary, nudge the table one way or the other until the angle is exact. Adjust the tilt scale so the indication mark on the table lines up with the zero line on the scale.

On some drill presses, the accuracy of the primary alignment setting will depend on the manufacturing quality: The indexing pin that passes through the table's apron and into a hole that is in the table bracket (the component that moves up and down on the column) determines the table's squareness. Check the procedure in the owner's manual.

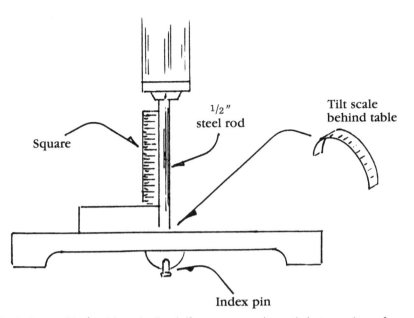

Square

1/2" steel rod

Tilt scale behind table

Index pin

5-1   Check the machine's table and adjust it if necessary so the angle between its surface and the centerline of the spindle is 90 degrees.

STORAGE PROJECTS

You can add to operational convenience by making a column storage unit that will hold frequently used tools within easy reach (FIG. 5-2). The idea is to avoid having to move from the tool to a cabinet or storage bench everytime you need a cutter or must use the chuck key. The split clamp locks for the unit can be positioned anywhere between table and headstock and are positioned so the swiveling shelf can be easily moved.

To make the split clamps, first form holes that equal the diameter of the column (more on this later in the chapter). Then saw them apart on their centerlines. Construction details for the project are offered in FIG. 5-3. Check your own needs before you drill the holes in the shelf.

The column-mounted support arm in FIG. 5-4 can be used with storage units of different design. The arm in FIG. 5-5 also grips the column by means of a split clamp, but in this case, half of the component is hinged so that locking it requires only a single bolt and nut.

The design for the storage unit to attach to the arm is a personal choice. Figure 5-6 suggests that they can be small cabinets or open shelves. Don't make them too large; their purpose is to hold frequently used tools, not a total assortment of drill press cutters.

The arm that was designed for storage units can also serve to support home-made, oversize tables (FIG. 5-7). An important feature is that the table will be held securely when in use, and can be swung aside when it is not needed.

Another idea for storage is to attach a utility tray to the back edge of the drill press table (FIG. 5-8). This requires drilling two holes through the table ledge for the 1/4-inch attachment bolts.

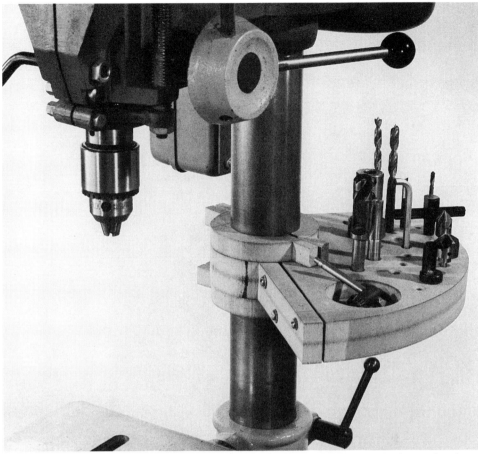

5-2 A column-mounted rack is handy for keeping frequently used tools close at hand.

LAYOUT

The first step in any drilling operation, and the most important one, is to accurately mark the hole location. A common and practical procedure is to draw intersecting lines—the intersection showing the point where the drill bit must enter. Because it's often difficult, especially with twist drills, to set the bit's point directly on the intersection, many careful operators adopt the procedure shown in FIG. 5-9. A steel rod with a sharp point that makes it easy to target the intersection is gripped in the chuck. The work is then clamped, and the drill bit is substituted for the rod. It's a no-fail method that is often recommended for metal drilling, but it's just as applicable for wood and other materials.

LAYOUT TOOLS

Figure 5-10 shows an assortment of layout tools that actually are, or should be, commonplace in a woodworking shop. Because you depend on items like this for accuracy, it pays to buy quality products.

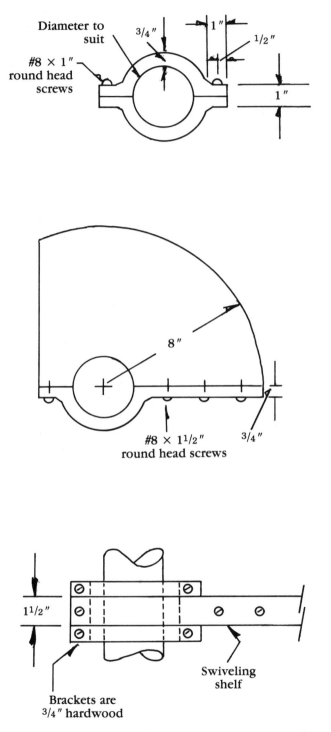

5-3 Construction details for the column-mounted storage unit.

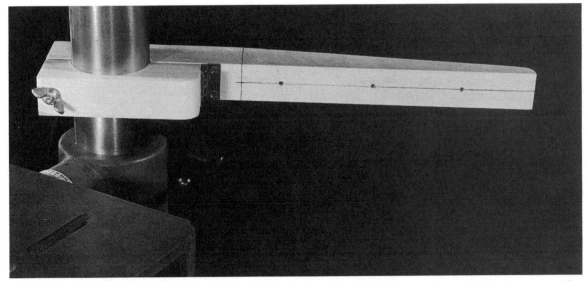

5-4 An arm that is designed like this can support various types of storage units as well as table jigs that are used for drilling applications.

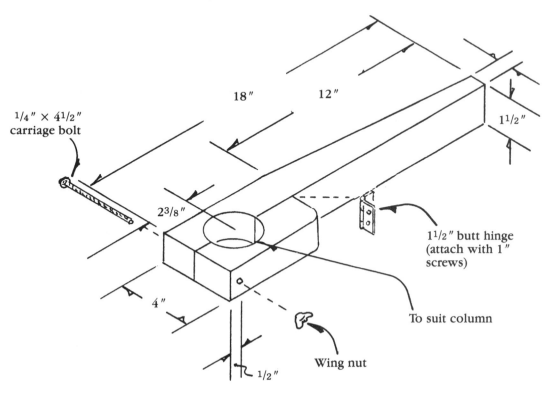

5-5 Construction details for the arm.

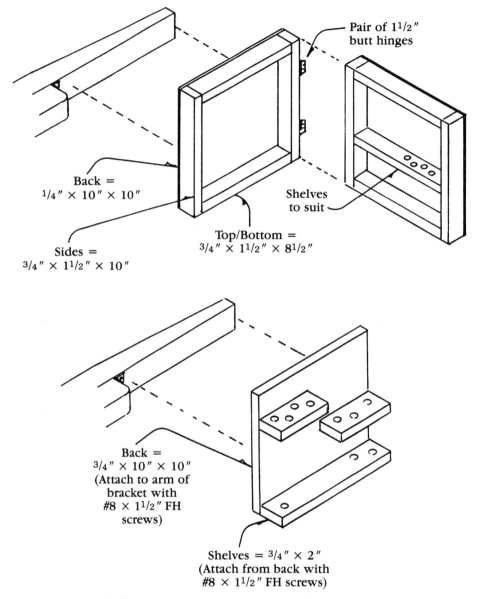

Pair of 1¹/2"
butt hinges

Back =
¹/4" × 10" × 10"

Shelves
to suit

Sides =
3/4" × 1¹/2" × 10"

Top/Bottom =
3/4" × 1¹/2" × 8¹/2"

Back =
3/4" × 10" × 10"
(Attach to arm of
bracket with
#8 × 1¹/2" FH
screws)

Shelves = 3/4" × 2"
(Attach from back with
#8 × 1¹/2" FH screws)

5-6 Typical storage units that can be attached to the arm.

The popular combination square is a layout tool that serves many purposes
(FIG. 5-11). Determining 90-degree and 45-degree angles is a snap (FIG. 5-12), and
because its blade is removable, it can be used like an ordinary ruler. The tool is
commonly used to mark lines parallel to an edge. Lock the blade at the edge distance required and then move the head along the work edge as you follow the
outboard edge of the blade with a pencil (FIG. 5-13). You'll get similar results if you
use the square to mark points at opposite ends of the workpiece and then connect
the points with a straightedge.

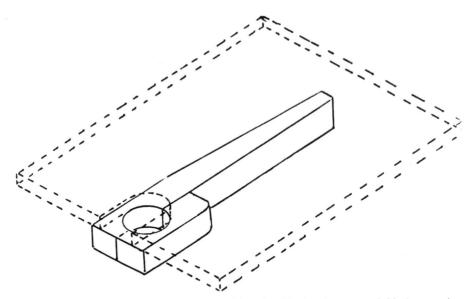

5-7 The arm can be used to support auxiliary tables. A table that is mounted this way can be swung aside and used as a storage shelf when it is not needed for drilling applications.

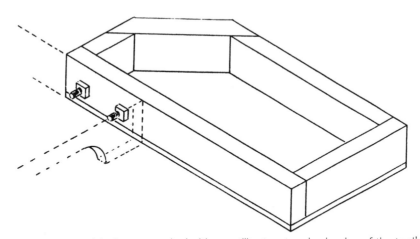

5-8 You can also provide for storage by bolting a utility tray to a back edge of the tool's table.

A common layout error when using a scale is to lay it flat and then scratch the dimension point (FIG. 5-14). The resulting broad line will not be accurate. It's better to tilt the scale or hold it vertically and then mark the location by making a small dot. When using a scale that has grooves at graduation points, use them to guide the point of the pencil, or scriber if you are using one (FIG. 5-15).

Always work with a hard pencil, at least 3H, and keep it sharp. A carpenter's pencil is acceptable for drawing lines if you maintain its edge as if it were the cutting edge of a chisel, but don't use one to mark location dots. Draw lines lightly so

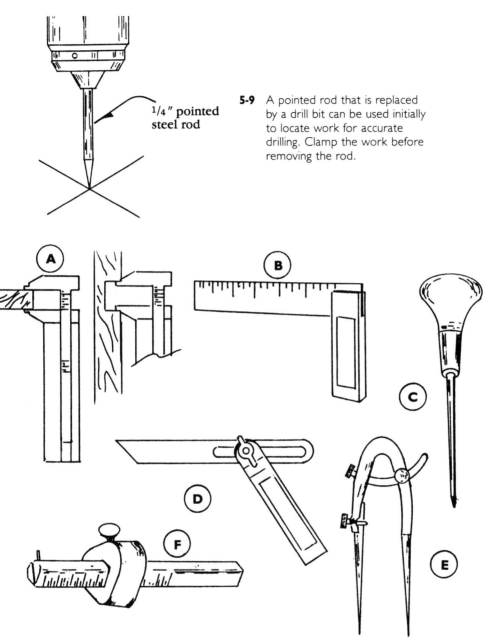

¹/4 " pointed
steel rod

5-9 A pointed rod that is replaced by a drill bit can be used initially to locate work for accurate drilling. Clamp the work before removing the rod.

5-10 An assortment of tools that are handy for doing layout work:
(A) Calipers for checking inside and outside dimensions
(B) Rule for measuring, checking squareness of edges, drawing lines parallel to edges
(C) Awl for scribing lines, indenting for small wood screws
(D) T-bevel for checking, marking various angles
(E) Dividers for drawing arcs or circles, spacing holes, marking lines parallel to edges, tracing profiles
(F) Marking gauge for marking lines parallel to edges

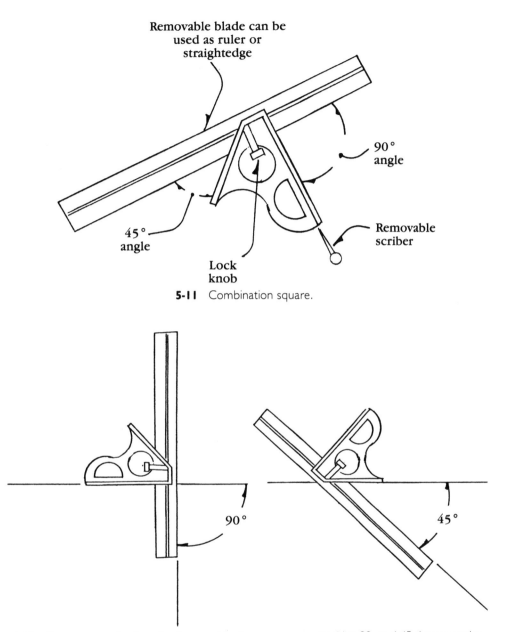

Removable blade can be
used as ruler or
straightedge

90°
angle

45°
angle

Removable
scriber

Lock
knob

5-11   Combination square.

90°

45°

5-12   Common applications for the combination square are marking 90- and 45-degree angles.

they can be easily erased. It's a nuisance to have to do sanding later to remove
layout marks.

When matching holes are needed in the edges of multiple pieces, say, for
edge-to-edge dowel joinery, do the layout on a single piece. Then, holding the
parts together—preferably with a clamp—use a square to carry the lines across all
pieces (FIG. 5-16).

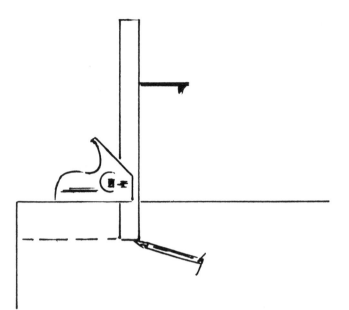

5-13 Using the square to draw a line parallel to an edge. Move square and pencil in unison.

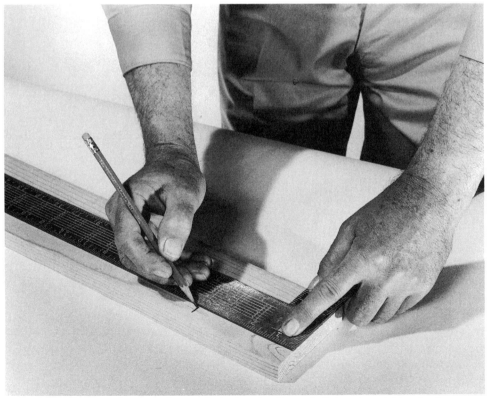

5-14 Don't make dimension marks by keeping the scale flat and scratching the line.

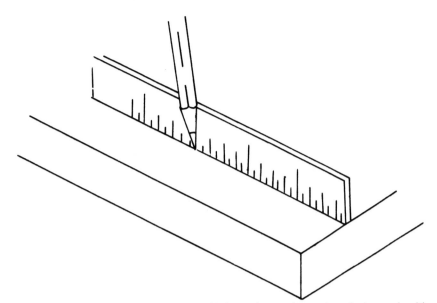

5-15 You will work more accurately if you hold the scale on edge and mark the work with a dot. Use a hard pencil for layout work and keep it sharp.

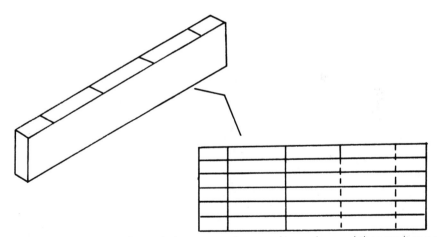

5-16 When the same layout is needed on many edges, mark one piece and then use it as a guide to mark all the others. Hold the pieces firmly together, preferably with a clamp.

Dividers or a compass can be used for more than drawing circles or arcs. They are useful for picking up a dimension from one component, or from a drawing, so it can be established on another part, and for making the locations of equally spaced holes (FIG. 5-17). The points of the tool are set for the spacing, then it is simply swung from one point to mark the following one. This method is a lot more accurate than marking spacing with a ruler and pencil.

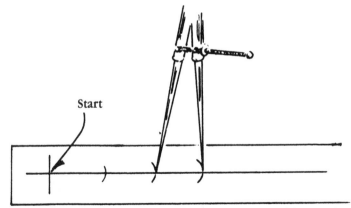

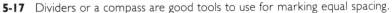

Start

5-17   Dividers or a compass are good tools to use for marking equal spacing.

DOWEL CENTERS

Dowel centers are handy little items that make it easy to accurately locate holes in mating pieces that will be connected with dowels (FIG. 5-18). The two components are pressed together after the centers have been inserted in holes drilled in one of the pieces (FIG. 5-19). The points on the centers form an indentation that tell you precisely where to drill so the holes will mate perfectly.

Locating centers

Some combination squares include a V-shaped component to find the center of round or square pieces (FIG. 5-20). The work is snugged in the V and a line is

5-18   Dowel centers are usually available in sets.

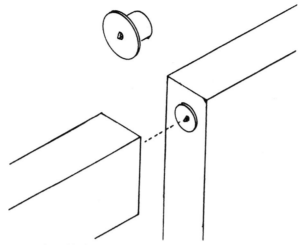

5-19 Dowel centers are placed in holes drilled in one part. When the mating piece is pressed in place, correct hole locations will be marked by the points on the centers.

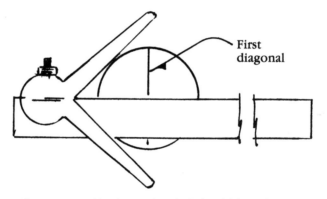

First diagonal

5-20 Components for some combination squares include a V-shaped head that is used as a center finder. A second line is drawn approximately at right angles to the first one. The intersection of the lines tells the exact center.

marked by following the edge of the blade. A second line is marked after the work has been turned about 90 degrees. The center is where the two lines intersect.

Figure 5-21 offers a plan for making a center finder that works as well and is as accurate as a commercial one. The project can be made larger for bigger pieces of material.

Figure 5-22 shows how to find centers on uniform pieces and on irregular shapes where a center finder isn't applicable. On squares and rectangles, simply draw lines from opposite corners. The center will be where the lines cross. On irregular pieces, use a compass to mark lines parallel to all edges. The center will be within the confined area established by the lines. Then you can determine (not precisely, but close enough) where the center is.

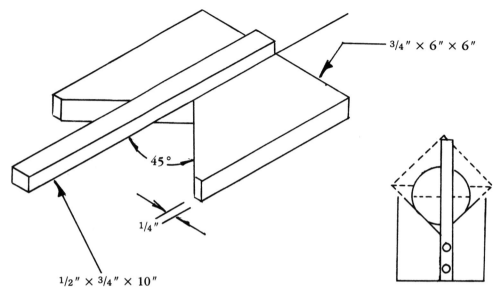

3/4" × 6" × 6"

45°

1/4"

1/2" × 3/4" × 10"

5-21 A center finder that you can make. Tools like this can be used to find the center of square as well as round stock.

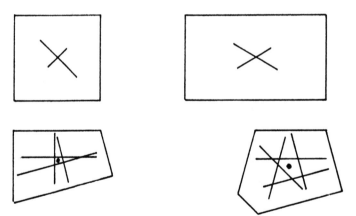

5-22 Diagonal lines locate the center of square, rectangular, and irregular pieces. On irregular pieces, mark lines drawn parallel to edges, then approximate the center.

Figure 5-23 suggests one way to mark an accurate, longitudinal centerline on a cylinder. Place the part on a flat surface and against a straight piece of wood. Hold the two pieces together and follow the edge of the wood strip with a pencil.

SPEEDS AND FEED

Speed refers to the RPM of the cutting tool; *feed* applies to how fast the cutter penetrates the work—which relates to the force applied on the feed lever, or, in the case of operations like routing and shaping, how fast the work moves past the cutter.

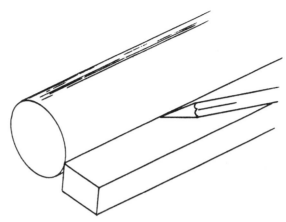

5-23 How to draw a longitudinal centerline on a cylinder or tube. The thickness of the guide strip does not matter.

Excessive speed can be dangerous on some tools, on others it can result in burn marks on both tool and work. The general rule is to use lower speeds with larger tools or, when in doubt, to start at a low speed and increase it until the center is functioning efficiently. Be aware that the optimum setup allows the tool to cut steadily with reasonable pressure on the feed lever (FIG. 5-24).

5-24 Drilling tools must cut smoothly and consistently without excessive pressure on the feed lever.

Speeds that are too slow or a feed that is overly cautious are also poor practice. Allowing a tool to just run instead of cut won't accomplish much and will cause quick dulling of cutting edges. A combination of slow speed and heavy feed can cause the cutting tool to snag, which—if the chuck doesn't just spin around the shank of the tool—will stall the motor and possibly cause damage to the tool or the work, or both.

Warning signs to heed that might indicate the wrong feed or speed, or both, include: rough results, unusual noises, excessive vibration, chatter, and excessive slowing or stalling of the motor.

Speeds for various hole sizes in soft and hard species of wood are listed in TABLE 5-1. Accept these as starting points. There are too many factors that affect the ideal speed, among them the bit, whether drilling is done into or across the wood grain, and the density of the wood (even boards cut from the same tree can differ in hardness). If the starting speed is not satisfactory, move up or down a bit until results are acceptable.

Table 5-1 Suggested Spindle Speeds for Drilling Wood

HOLE SIZE	SPEED (RPM) Softwood	Hardwood
1/16 "	4700	4700
1/8 "	4700	4700
3/16 "	4700	2400
1/4 "	2400	2400
5/16 "	2400	1250
3/8 "	2400	1250
7/16 "	2400	1250
1/2 "	1250	1250
5/8 "	1250	700
3/4 "	1250	700
7/8 "	1250	700
1 "	700	700
1 1/4 "		
1 1/2 "	USE LOWEST SPEED	
2 "		

BACKUP AND WORK SECURITY

Backup has to do with placing the work on a piece of scrap material when you are drilling through holes. Without one, you will discover excessive splintering and feathering when you examine the underside of the workpiece, and this will be especially true when drilling materials like plywood and particleboard (FIG. 5-25). If the appearance of the underside of the work is not critical, you can place the work directly on the table and drill through, but be sure the opening in the table is aligned with the spindle. In any case, using a backup is good procedure.

Work security has to do with being sure the workpiece remains where it should throughout the drilling process. Because the work has a tendency to twist

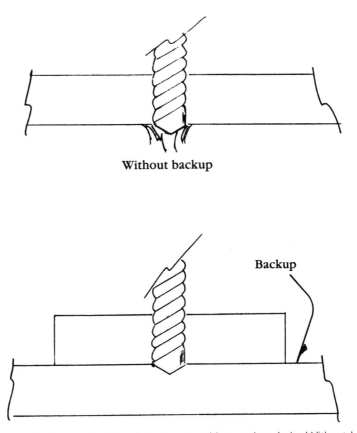

Without backup

Backup

5-25 Always use a backup under the workpiece to achieve a clean hole. Without backup, the work will splinter when the bit breaks through.

along with the rotation of the bit, keeping it still is important for safety as well as accuracy. The inside edges of long pieces can be braced against the left side of the column to counteract twist, but clamping is the best procedure. Clamping will also guarantee that the work will remain on the table anytime you retract the cutter.

Commercial clamps (FIG. 5-26) are available, but common workshop clamps will do as well. C-clamps are especially useful and they will not cause any damage if you place a small pad between the work and the clamp. Screw clamps do a good job, especially when the work must be held on edge (FIG. 5-27), and they, in turn, can lock to the table with a C-clamp. Thus you have work security plus assurance that the workpiece will be held in true vertical position.

The only time you might ignore the clamping rule is when the workpiece is large and small drills with negligible twist action are used. Always clamp small pieces, regardless of how small the bit might be; there is no point in hand-holding work if it puts your fingers close to the cutting area.

Often, when a fence is used as a drill guide—as when forming a series of holes that have a common edge distance—you can get by without clamps. Any

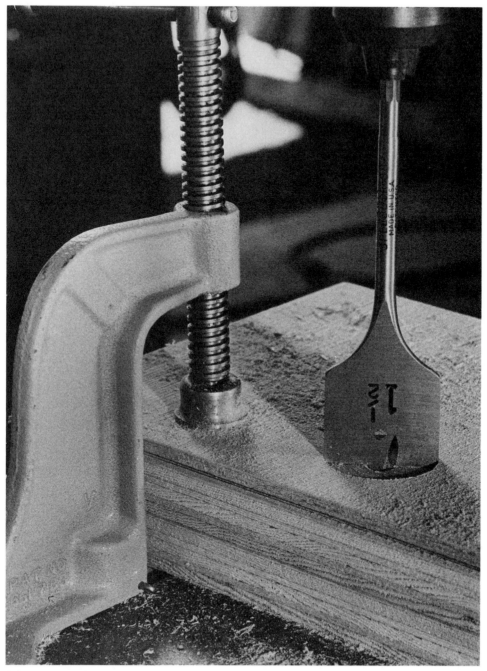

5-26   A type of clamp that is designed specifically for drill press use. The bottom of the clamp has an integral bolt that passes through a slot in the table.

5-27 Handscrews are very good clamping devices for drill press work. They are especially useful for gripping parts that must be held on edge.

twisting force exerted by the bit will be opposed by the fence, not your hands. Fence designs for various applications are included in this chapter.

DRILLING STOPPED HOLES

Stopped or blind holes are those that penetrate the work only partway. You can use one of two systems to organize the drill press so you can form holes of predetermined depth.

First, mark the hole depth on an edge of the stock and then extend the bit to that point (FIG. 5-28). Lock the quill to hold the setting and then set the lock nut on the stop rod so the quill can be extended no farther. The second method is to bring the bit down until it contacts the surface of the work and then set the stop rod lock nut so it will prevent the quill from extending farther than the required hole depth (FIG. 5-29).

Methods to control hole depth without using the machine's stop rod involve small gadgets that are used on the bit. Figure 5-30 shows a sample that is adjustable for various bit diameters. Its distance from the end of the bit determines how deep you can drill. Stop collars (FIG. 5-31) work in similar fashion.

Once you have organized for the task, you can drill any number of holes knowing all of them will be equal in depth.

CORRECTIONS

You can use the idea shown in FIG. 5-32 if you have erred when marking a hole location or have inadvertently drilled an undersize or even an oversize hole. To

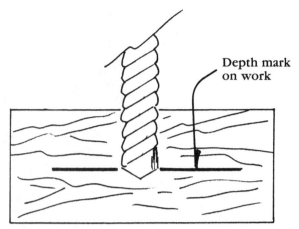

5-28 One way to establish hole penetration is to lock the quill when the bit is level with a mark on the work. Then, adjust the lock nut on the stop rod so the quill can be extended no farther.

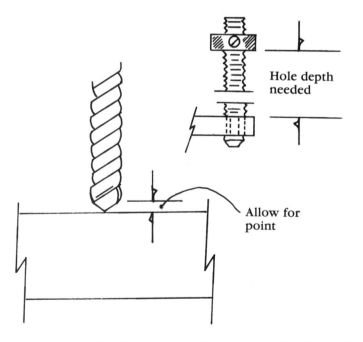

Depth mark on work

Hole depth needed

Allow for point

5-29 A second method for limiting bit penetration: Set the stop rod for the depth of the hole while the drill bit is contacting the surface of the work.

correct the diameter of an existing hole, plug it tightly with a dowel and, after marking its center, redrill for enlargement or reduction. To prevent the plug from twisting, coat it with glue and form the new hole after the glue has dried. Use the same idea to establish a new hole center.

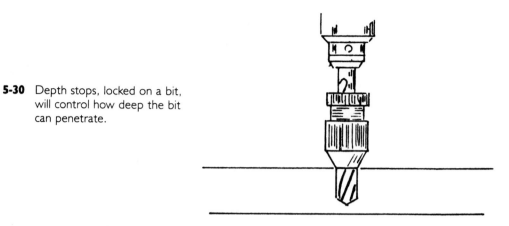

5-30 Depth stops, locked on a bit, will control how deep the bit can penetrate.

5-31 Stop collars lock on the bit by means of a set screw. The hole in the collar should match the diameter of the bit.

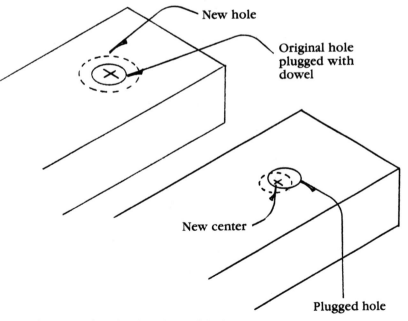

New hole

Original hole plugged with dowel

New center

Plugged hole

5-32 Enlarging holes or relocating them is possible if you use a dowel to plug the original hole. The dowel makes it possible to establish a new center.

A BASIC FENCE

A fence that can be clamped to the table makes it easy to do many drilling jobs with accuracy—for example, forming a series of holes on a common centerline or controlling the edge distance of similar holes (FIG. 5-33). The fence, detailed in FIG. 5-34, is not an elaborate project. However, make it carefully so that when it is in place its bearing surface will be square to the machine's table. This same fence, with some modifications, will be shown in this chapter serving a variety of applications.

Hardware

There are many jig designs presented in this book, some of them requiring particular hardware.

- *Carriage Bolts* are oval-head fasteners with shoulders that dig into the wood to keep the shaft from turning when the nut is tightened (FIG. 5-35).

5-33 The basic fence in use. Here it is controlling edge distance and assuring that the holes will be on a common centerline.

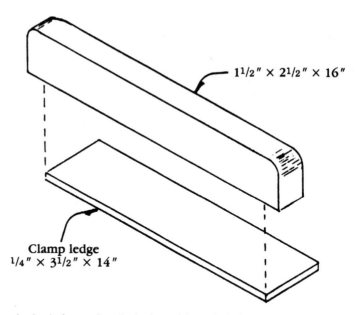

$1^{1}/_{2}'' \times 2^{1}/_{2}'' \times 16''$

Clamp ledge
$^{1}/_{4}'' \times 3^{1}/_{2}'' \times 14''$

5-34 Making the basic fence. Attach the base (clamp ledge) with glue and flat-head screws.

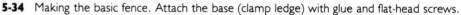

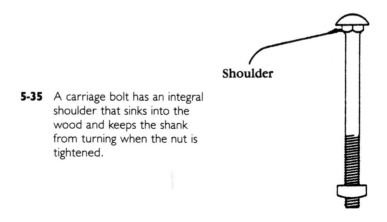

Shoulder

5-35 A carriage bolt has an integral shoulder that sinks into the wood and keeps the shank from turning when the nut is tightened.

This is convenient because you can do the tightening with a single wrench or socket, whereas two are needed with other types of bolts.

- *Threaded Sleeves and T-Nuts* (FIG. 5-36) make it possible to have metal threads in wood. Thus, components can be joined with a conventional bolt, and a bolt may be used as a locking device when part of a jig must be adjustable.

- *Bushings* (FIG. 5-37) are steel inserts that eliminate wear and tear on a hole when, for example, a drill bit must be frequently retracted or the position of a guide rod must be accurate. The products are available in various materials and in a wide assortment of lengths and inside/outside diameters.

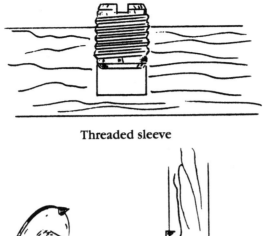

Threaded sleeve

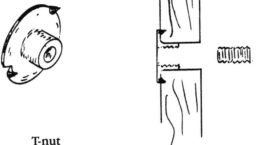

T-nut

5-36   Threaded sleeves and T-nuts are used to supply metal threads in wood.

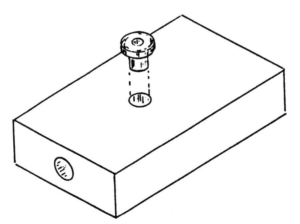

5-37   Bushings eliminate wear and tear on a hole that is used as a guide for drilling.

JIGS

Some workers make a distinction between a ''jig'' and an ''accessory,'' and there is justification for this. Although the terms can be used interchangeably, here a jig will refer to a user-made device that either extends the tool's applications or makes it easier to assure accuracy, while an accessory will refer to a user-made drum sander or spot polisher.

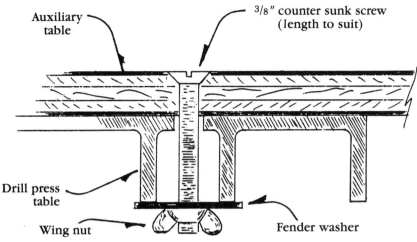

5-38 This fastening method can be used instead of clamps for securing auxiliary tables.

A common start in the jig area is to supply an extra-large worktable. This can simply be a handy piece of plywood clamped in place. However, because an additional work surface is often required, it's best to select a piece of cabinet grade plywood and to attach it along the lines shown in FIG. 5-38. Thus, the unit will be a permanent auxiliary table that can be quickly put in place or removed.

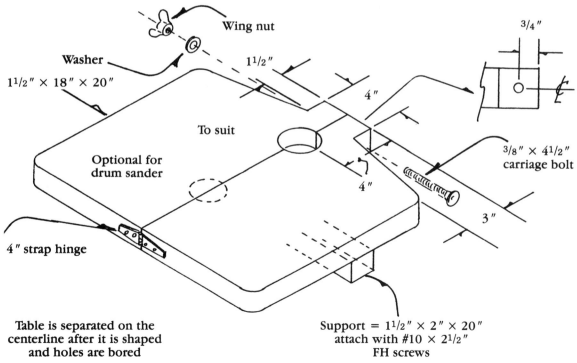

5-39 Auxiliary table that is designed like a split clamp. The support is secured to only one half of the table.

Extra tables can be more sophisticated and more functional. The one detailed in FIG. 5-39 is designed like a split clamp. Additional convenience is created because the table can be swung aside when it is not needed. In its off-side position it can serve as a shelf for drilling tools.

Using the split clamp design to secure jigs is generally acceptable. There are operations, however, like routing and shaping, where the work is moved against a cutter, creating a tendency for the jig to swivel. In such cases, it's wise to use the system shown in FIG. 5-40 instead of a split clamp. The lock bolt, with a pad of hardwood placed between its end and the drill press column, will snug the jig in place.

An auxiliary table can also be equipped with an adjustable fence. The added feature comes in handy for operations like gauging edge distance of holes, and drilling on a common centerline (FIG. 5-41).

An ultimate version of table design is shown in FIG. 5-42. It offers not only an adjustable fence, this time one that is clamped in place, but also a storage drawer. The U-shaped interior of the drawer provides access for a drum sander that passes through the hole of the table. Drum sanding operations will be covered in Chapter 11.

Jigs that you design

There are jigs that become permanent shop equipment because they have universal applications, and others that are for a specific, possibly one-time use. Generally, the latter type are used to facilitate a drilling procedure and guarantee accuracy when the hole is required on similar pieces. The jig must be specially designed for such production work. In this particular area *you* are the creator. Some example units that will help you become jig-wise are shown in FIGS. 5-43 through 5-49.

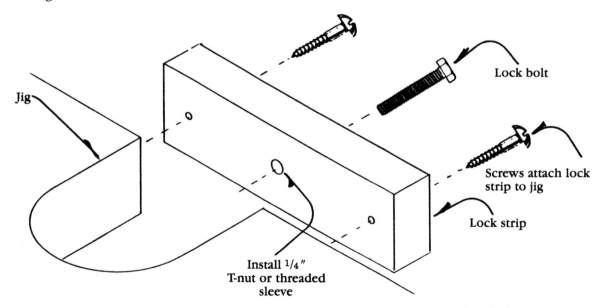

5-40 Clamping device for auxiliary tables and jigs that can be used in place of a split clamp.

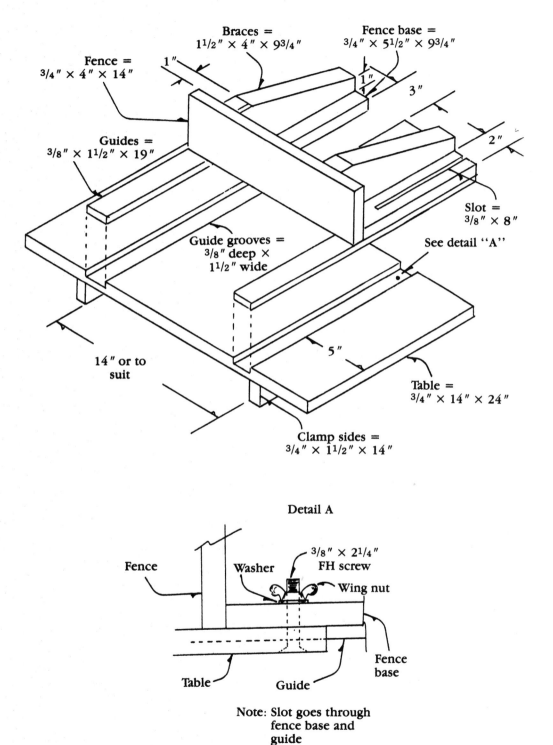

Braces =
1½″ × 4″ × 9¾″

Fence base =
¾″ × 5½″ × 9¾″

Fence =
¾″ × 4″ × 14″

1″

1″

3″

2″

Guides =
⅜″ × 1½″ × 19″

Slot =
⅜″ × 8″

Guide grooves =
⅜″ deep ×
1½″ wide

See detail "A"

14″ or to
suit

5″

Table =
¾″ × 14″ × 24″

Clamp sides =
¾″ × 1½″ × 14″

Detail A

Fence

Washer

⅜″ × 2¼″
FH screw

Wing nut

Fence
base

Table

Guide

Note: Slot goes through
fence base and
guide

5-41 Plans for an auxiliary table with an adjustable fence.

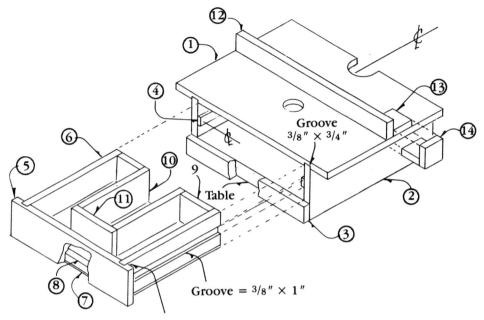

Groove
3/8" × 3/4"

Table

Groove = 3/8" × 1"

Dado = 3/8" × 3/4"

Materials List

Key	Part	#Pcs.	Size	Material
Case				
1	Table	1	$3/4" \times 15" \times 20"$	Cabinet grade plywood
2	Sides	2	$3/4" \times 5^7/8" \times 15"$	"
3	Clamp ledge	1	$3/4" \times 1^1/2" \times 15^1/2"$	Hardwood
4	Drawer guides	2	$3/8" \times 1" \times 15"$	"
Drawer				
5	Front	1	$3/4" \times 4" \times 15^1/2"$	Plywood
6	Sides	2	$3/4" \times 3^3/4" \times 15^3/8"$	"
7	Bottom	1	$1/4" \times 14" \times 15"$	"
8	Cleat	1	$3/4" \times 3/4" \times 12"$	Hardwood
9	Back	2	$3/4" \times 3^3/4" \times 4"$	"
10	Compartment sides	2	$3/4" \times 3^3/4" \times 9"$	"
11	Compartment front	1	$3/4" \times 3^3/4" \times 5^1/2"$	"
Fence				
12	Fence	1	$3/4" \times 2" \times 20"$	Hardwood
13	Guide/Clamp	2	$3/4" \times 2^1/2" \times 2^1/2"$	"
14	" "	1	$3/4" \times 2^1/4" \times 2^1/2"$	"

5-42 Plans and materials list for auxiliary table, including storage drawer. Check the dimensions to be sure they are suitable for your machine.

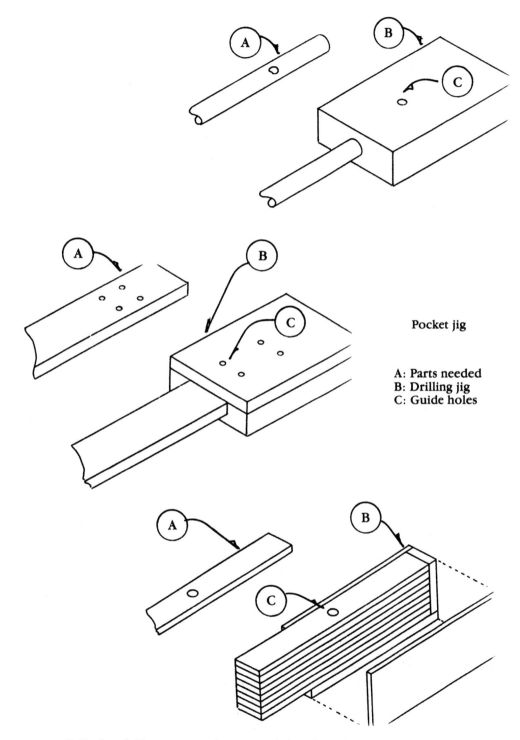

Pocket jig

A: Parts needed
B: Drilling jig
C: Guide holes

5-43 Jigs of this nature must be custom designed to suit particular applications.

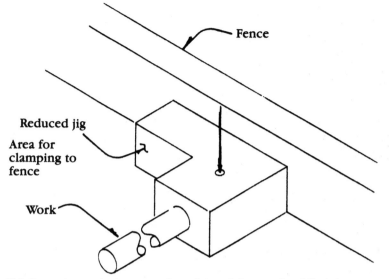

5-44 This jig can be used when a number of dowels have to be drilled the same way.

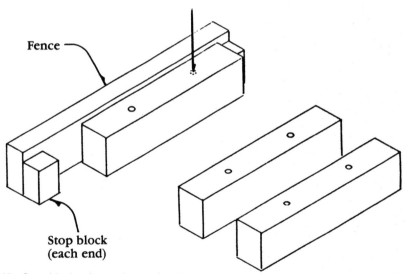

5-45 Stop blocks clamped or tack-nailed to a fence often serve as hole location jigs.

AUTOMATIC HOLE SPACING

It's often necessary to drill a series of equally spaced holes on a common centerline. If you do this by layout and pencil mark hole locations, accuracy will depend on variable factors. This can be eliminated by using a hole spacing jig (FIG. 5-50). The jig is made by drilling a series of holes on the centerline of a bracket that is secured with screws to the top edge of the basic fence.

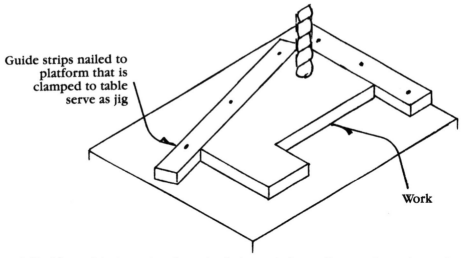

Guide strips nailed to
platform that is
clamped to table
serve as jig

Work

5-46 It's possible that strips of wood nailed to a platform will serve adequately as a jig.

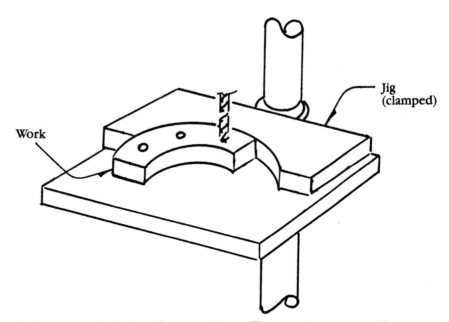

Jig
(clamped)

Work

5-47 Jigs can also be designed for curved pieces. The curves in project and jig must match.

Drill one hole in the stock and while the bit is still in place, clamp the fence to provide the necessary edge distance (the distance from the center of the hole to the edge of the board). The distance between the guide pin and hole will equal the required spacing. The guide pin is used to engage a drilled hole so it positions the work for the next one. The diameters of the guide pin and bit must match, but they can be used for pilot holes that will later be enlarged. You can, of course,

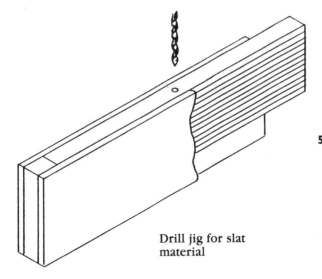

5-48   A typical jig for drilling slats.

Drill jig for slat material

5-49   A slat jig in use. Use a clamp to supply extra security for the parts.

5-50 This hole spacing jig has a bracket that is attached to the basic fence. Fence and bracket work together to establish edge distance and hole spacing.

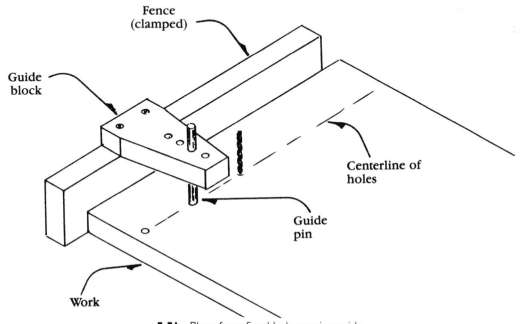

**Fence
(clamped)**

**Guide
block**

**Centerline of
holes**

**Guide
pin**

Work

5-51 Plans for a fixed hole spacing guide.

work for finished holes. If they must be ¹/₄ inch, then drill the bracket for a
¹/₄-inch guide pin and use a bit of similar diameter. Details for the jig are shown in
FIG. 5-51.

Figure 5-52 provides construction details for making a more flexible auto
spacing jig. An arm that can be rotated, moved to-and-fro, and set at infinitely variable heights is used in place of a fixed bracket. Thus, the guide pin can be located
for spacing by adjusting the arm after the fence has been positioned for edge distance. Because the project incorporates a bushing and specifies a ¹/₈-inch diameter
guide pin, it is used for drilling pilot holes that can then be enlarged to necessary
size (FIG. 5-53). Being able to raise the arm allows using the jig for equally-spaced
holes in stock edges as well as into surfaces (FIG. 5-54).

Another gauge that can be made for use with the basic fence is detailed in FIG.
5-55. The clamps that secure the inboard dowels are screwed to the back of the
fence at each end so the gauge can be situated either left or right. By adjusting the
holding block and the dowels you can establish a particular distance from dowel
to drill bit. A typical use for the gauge is drilling a hole on multiple pieces a certain
number of inches from the end of the stock. If you have ever used a table saw you

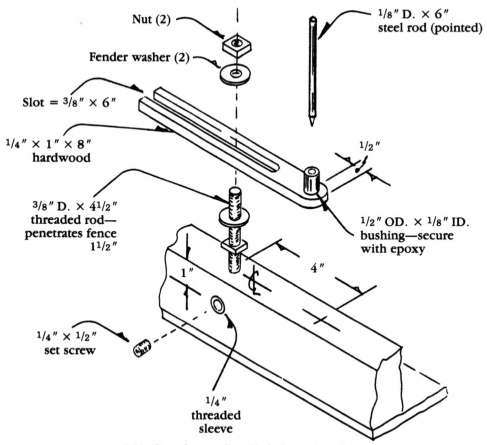

5-52 Plans for an adjustable hole spacing guide.

5-53   The adjustable guide provides greater flexibility in hole spacing.

5-54   Because the arm on the guide is adjustable for height, the jig may be used for auto spacing holes into edges as well as surfaces.

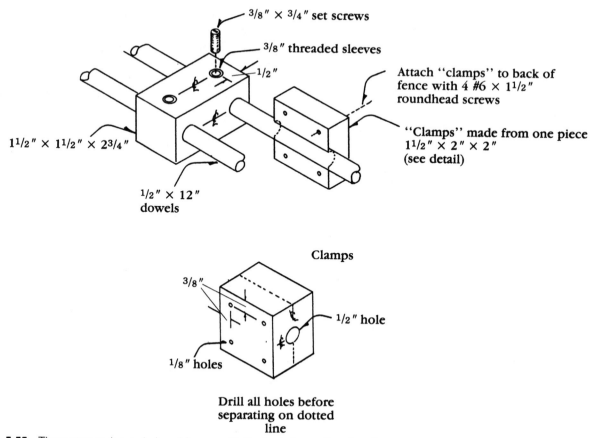

3/8" × 3/4" set screws

3/8" threaded sleeves

1/2"

Attach "clamps" to back of fence with 4 #6 × 1 1/2" roundhead screws

"Clamps" made from one piece 1 1/2" × 2" × 2" (see detail)

1 1/2" × 1 1/2" × 2 3/4"

1/2" × 12" dowels

Clamps

3/8"

1/2" hole

1/8" holes

Drill all holes before separating on dotted line

5-55 These stop rods are designed for use with the basic fence. One function is to gauge the distance of a hole from the end of a workpiece.

will recognize that the concept is based on stop rods that are used with a miter gauge.

DRILLING DEEP HOLES

The deepest hole you can drill without relying on special arrangements is determined by the maximum extension of the quill. If the limit is 4 inches, that's as far as you can go—even if the bit in use is fluted for a 5- or 6-inch depth. One solution that will allow twice the penetration is to drill pilot holes on a common, vertical centerline into opposite surfaces of the material. Then, after drilling into one surface, the stock is inverted and a second hole is formed to meet with the first one (FIG. 5-56). The critical accuracy factor is the layout for the pilot holes.

To avoid the layout chore, you can use the method shown in FIG. 5-57. The clamped strips of wood are positioned so that the center of the workpiece will be under the point of the bit. The jig will assure that the opposing holes will meet as they should.

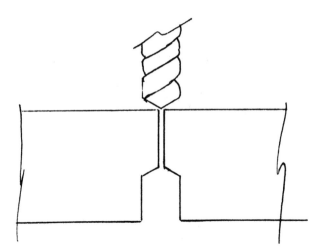

5-56 You can double how far quill extension allows you to drill by boring from both sides of the stock. An initial pilot hole serves as a guide.

5-57 Alignment of opposing holes can be assured by using clamped strips of wood as a jig.

For greater assurance of accuracy, especially when the drilling chore is needed for similar parts of a project, you can use the technique shown in FIG. 5-58. The work is automatically positioned correctly for the second hole because of the guide pin that is of the same diameter and concentrically aligned with the drill bit. After the first holes are drilled in the parts, the guide pin, mounted on a simple platform, is clamped to the table so the opposing holes can be formed (FIG. 5-59).

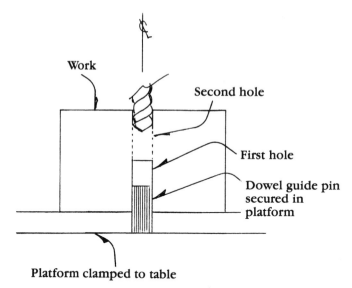

Work

Second hole

First hole

Dowel guide pin
secured in
platform

Platform clamped to table

5-58 Dowel jig can be used for drilling extra-deep holes.

5-59 The jig is clamped in place so the dowel guide pin (arrow) is aligned with the bit. Pin and bit must have the same diameter.

The diameters of guide pin and bit can suit the size of the project hole, or they can be much smaller so as to produce pilot holes that can be enlarged later to the required size. The latter idea is practical because the jig can then be maintained for use in various situations.

Extension bits can be used by adopting a procedure that gets around the limitations of quill extension. Drill the first hole as deep as you can. Then, after shutting down the machine, raise the table so the bit will bottom in the existing hole and drill again. Obviously, the idea is more applicable to a floor model machine because of its chuck-to-table (or base) capacity.

Extension bits, because of their length, can whip, so use them with care. Use a slow speed, at least to start, and have the point of the bit bearing against the work before you flick the on switch. When drilling deep holes, retract the bit more often than you do normally. This will keep heat buildup on tool and work to a minimum; deep holes can capture a lot of waste.

DRILLING LARGE HOLES

Holes up to about $1^1/_2$ inches in diameter can be cut with tools like spade bits. Beyond this size, the work is usually done with special tools like hole saws and fly cutters.

Hole saws, like the one in use in FIG. 5-60, are available in individual units that produce a hole of specific size. The smaller the tool, the more RPM you can use.

5-60   Hole saws make cuts of specific diameter. Size range is from about 1 inch to 3 inches. The center pilot drills are replaceable.

However, as with any drill, it's good practice to start at the machine's lowest speed and to go higher only if the cutting goes smoothly and without vibration or screeching noises. Use a feed pressure that will keep the saw teeth working. Otherwise they will scrape and not cut, creating excessive heat that can harm tool and work. The common maximum size of hole saws is about 3 inches.

The advantage of a fly cutter is that it can produce infinitely variable hole diameters within its capacity (FIGS. 5-61 and 5-62). The cutting bits are easy to keep sharp and they can be ground for thin or wide cutting grooves. They can even be modified (as shown in Chapter 13) to do some amount of decorative surface carving.

The product shown in FIG. 5-63 is a newcomer in the fly cutter field. It has two cutting bits instead of one and a calibrated arm that is used to establish hole diameter. Two cutters equalize the load on the arm so the tool rotates more smoothly than a single-cutter version. Because the cutters are individually adjustable, you can offset one for a wider cutting groove or even to produce rings.

A bonus feature of hole saws and fly cutters is that they form holes by removing discs. Thus they can be viewed as tools for forming circular components like toy wheels and bases for projects.

Work with care when forming large holes, no matter what cutter is used. Use a suitable backup and enough clamps to keep the work secure. Keep your hands well away from the cutting area. Even at slow speed, the business end of a fly cut-

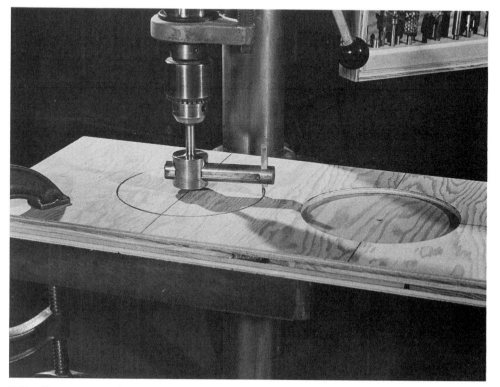

5-61 The arm on a fly cutter is adjustable so the tool can be used for any size hole within its capacity.

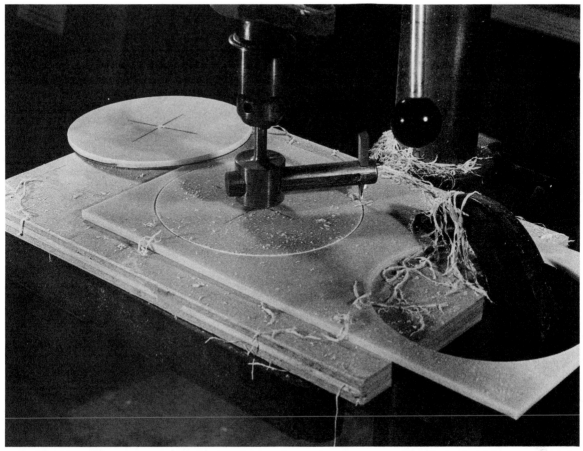

5-62 The bits on fly cutters are tempered so they can be used on many materials, Corian being just one example. Use tools like this with great care.

ter is just a blur, so bear that in mind before deciding where your hands should be. Actually, when the work is clamped in place, as it should be, your free hand doesn't have to be anywhere near the work area.

END DRILLING

End drilling involves forming holes in the end of components that are too long to be held on the table in the usual fashion. One solution is to swing the table aside and employ a handscrew, which in turn is clamped to the table as a gripping device for the workpiece (FIG. 5-64). The idea is practical but calls for care to be sure the work is in vertical alignment with the spindle.

Another, better way to go if the table design permits, is to tilt the table so it is parallel with the column (FIG. 5-65). This provides ample support for the work and it can, if necessary, be clamped in several places. A precaution to take, especially if you must drill into multiple pieces, is to provide a clamped fence that will assure vertical alignment with the drill bit.

5-63 The two-blade fly cutter is a newcomer. Two blades add to balance.

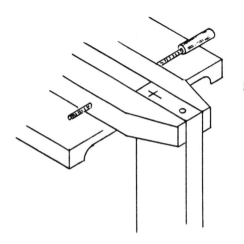

5-64 Using a handscrew to grip work for end drilling. Be sure the handscrew is clamped to the table and that the work is in vertical alignment with the spindle.

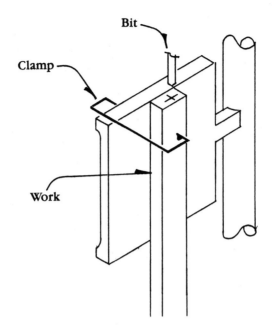

5-65 You can supply greater support for end drilling if the machine's table can be tilted like this.

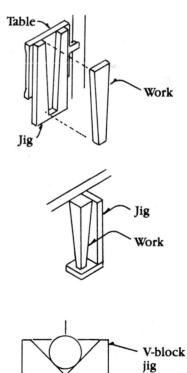

5-66 End drilling project components of particular shape will call for designing special jigs. These ideas will work only if the table is tiltable.

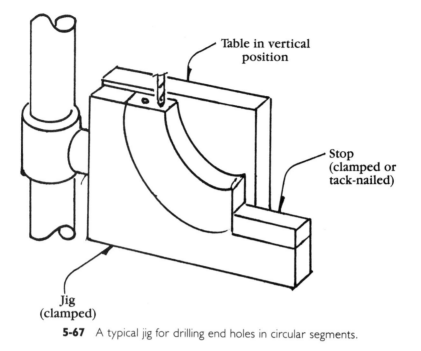

Table in vertical
position

Stop
(clamped or
tack-nailed)

Jig
(clamped)

5-67 A typical jig for drilling end holes in circular segments.

Special holding jigs are needed when the workpieces have an odd shape. These are designed to suit the component and are clamped to the table so the parts can be drilled accurately. Examples of typical jigs are shown in FIG. 5-66. Curved pieces can be handled the same way (FIG. 5-67). This is a good setup for segments that must be joined with dowels to form a ring.

Making a column jig

The column jig for end drilling is ideal for a floor model machine, but will also serve well for a bench model if the tool's base is used as a table (FIG. 5-68). The fence is adjustable to suit the thickness of the stock, which is clamped in place with its lower end resting on the machine's table or base (FIG. 5-69). When similar pieces must be drilled, it's a simple matter to clamp a strip of wood vertically to the fence so it will serve as a right angle stop.

When the jig is not being used for its designed purpose it can serve very nicely as a fence for jobs like drilling holes on a common centerline (FIG. 5-70). When used this way, the jig's fence rests directly on the tool's table. The jig is constructed as shown in FIG. 5-71. The guide rods must fit tightly in the holes drilled in the fence. For extra security, coat their ends with epoxy before tapping them into place.

CONCENTRIC DRILLING

Concentric drilling, which is really end drilling, can also be accomplished with the column jig by making a V-block so work can be clamped between block and fence (FIG. 5-72). The V-block can be simple and used with clamps, or it can be a

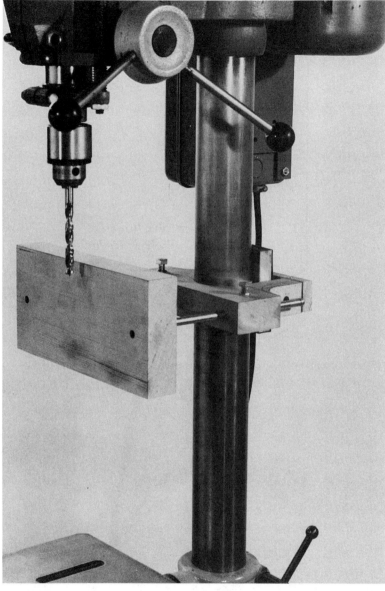

5-68 The column jig is designed primarily for end drilling. You must be sure that its fence will be on a vertical plane.

bit more sophisticated (FIG. 5-73). Here, fence and block are drilled for a pair of 1/4-inch bolts and wing nuts so a clamp is not needed to secure the work.

Positioning workpieces for concentric drilling can be accomplished in various ways. Jigs like those shown in FIG. 5-74 will provide assurance that workpieces will be held in true vertical position. It isn't necessary to get involved in constructing jigs of this nature unless there are many similar pieces to drill.

A good way to hold small pieces is to provide a special clamp by sawing matching Vs in the jaws of a handscrew. The device will be usable for square stock as well as round (FIG. 5-75).

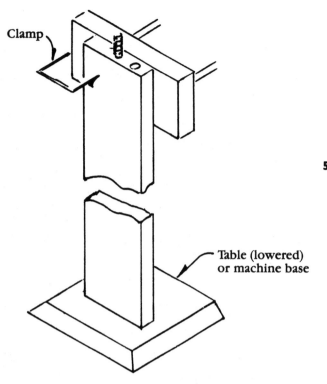

Clamp

Table (lowered)
or machine base

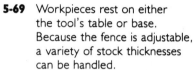

5-69 Workpieces rest on either
the tool's table or base.
Because the fence is adjustable,
a variety of stock thicknesses
can be handled.

5-70 The column jig may also be used as a routine fence. When used so, the jig's fence rests
directly on the tool's table.

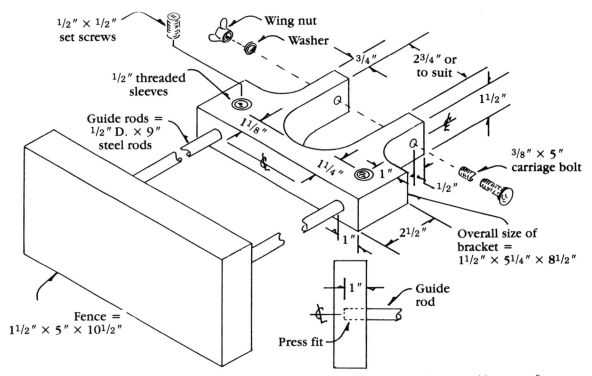

5-71 Construction details for the column jig. Be sure the cutout for the column provides a snug fit.

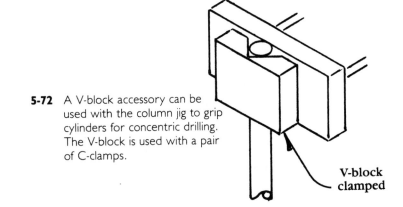

5-72 A V-block accessory can be used with the column jig to grip cylinders for concentric drilling. The V-block is used with a pair of C-clamps.

The V-block detailed in FIG. 5-76 and shown in use in FIG. 5-77, is designed for attachment to the basic fence. After the block is shaped, center it on the fence and clamp it in place so you can drill holes for the guide rods through both items. Like other jigs of this nature, it will hold square stock as well as cylinders.

5-73 This type of V-block becomes a permanent accessory for the column jig. Holes are drilled through it and the jig's fence for 1/4-inch bolts and wing nuts.

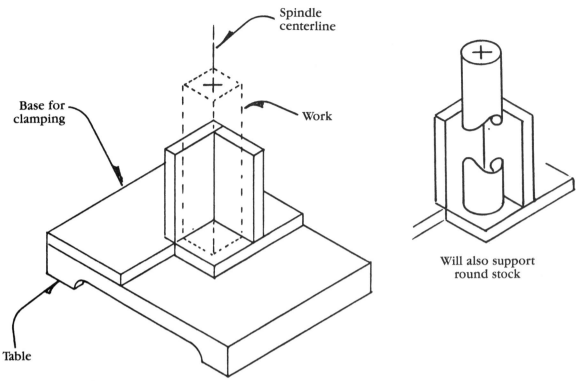

Spindle
centerline

Base for
clamping

Work

Table

Will also support
round stock

5-74 Other types of jigs you can make for holding cylinders or squares for concentric drilling.

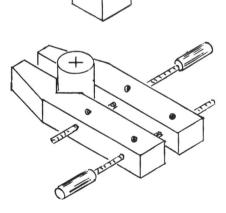

5-75 A modified handscrew will serve nicely as a vise for small parts. The Vs in the jaws will grip round or square pieces.

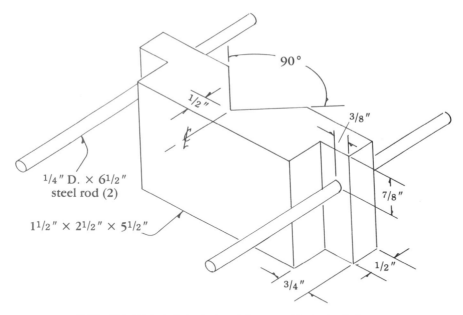

5-76   This V-block jig is designed for use with the basic fence.

90°

1/2"

3/8"

1/4" D. × 6 1/2"
steel rod (2)

7/8"

1 1/2" × 2 1/2" × 5 1/2"

1/2"

3/4"

5-77   The V-block position is established by the twin guide rods. Use a clamp to secure the work.

RADIAL HOLES

The term *radial holes* covers those that are drilled diametrically into or through a cylinder, those drilled on a circular layout, and those that are drilled into the edge of circular components.

Holes through cylinders or tubing require the use of V-block jigs that can be made in one of the two ways diagrammed in FIG. 5-78. The V-block is placed on the table so the bottom of the V is vertically aligned with the center of the drilling tool. Thus, the work that is snugged into the V will be drilled exactly on a diameter. When more than a single hole is needed the jig can be organized as shown in FIG. 5-79. The nail guide not only keeps the work from rotating so that holes will be on the same centerline, but it can be located to serve as a gauge for equally spaced holes. If used for the latter purpose, the nail guide should have its head filed off. V-blocks can also be used when it's necessary to drill diagonally through square stock (FIG. 5-80). Be careful when starting the hole to be sure the drill bit doesn't wander off the corner of the work.

Construction details for V-block jigs that are more universal are offered in FIGS. 5-81 and 5-82. In the first, the bridge-type guide block has a fixed position. The guide block for the second version is mounted on slides so the capacity of the jig is greatly increased. Both are used the same way (FIG. 5-83). After the first hole is drilled, the work is moved to be engaged by the guide nail—assuming, of course, that a series of equally-spaced holes is required. Use jigs like this to drill pilot holes that can later be enlarged. Enlarging should also be done with the jig, to be sure the new hole follows the path of the first one.

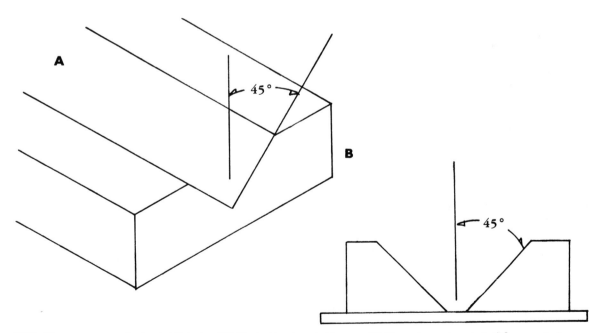

5-78 Two ways to make basic V-blocks: (A) Single V-cut on table saw. (B) Two pieces, beveled 45°, mounted on a base.

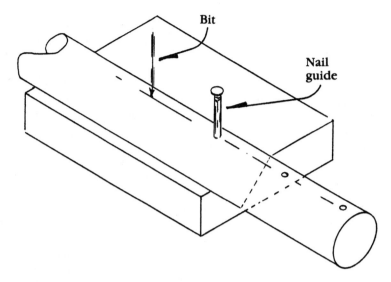

5-79   Any V-block can be used for hole spacing if you use a nail as a gauge. The nail engages the first hole and positions the work for the next one.

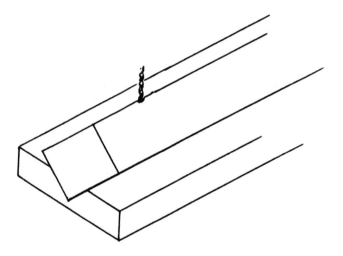

5-80   V-blocks can hold square pieces for diagonal drilling. Start carefully to avoid letting the bit wander off the corner.

V-blocks can solve some unique drilling chores. Figure 5-84 shows one that is set up for drilling spoke holes in a small wheel. The jig in FIG. 5-85 has intersecting Vs so it can cradle ball shapes.

EDGE HOLES

Drilling radial holes into edges requires a different arrangement and a different kind of V-block (FIG. 5-86). Tilt the table so its surface is parallel with the column

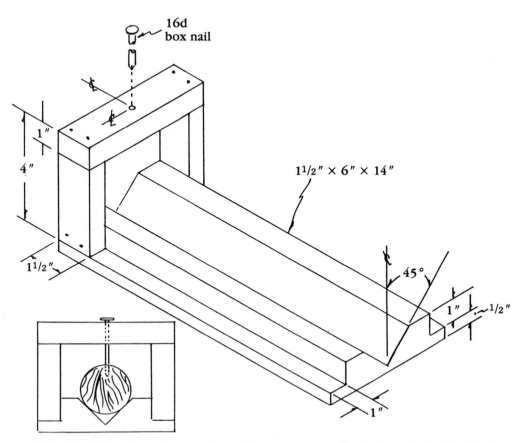

16d
box nail

1″

4″

1¹/₂″ × 6″ × 14″

1¹/₂″

45°

1″   1/2″

1″

5-81 This V-block jig is designed for variable hole spacing. The holes, drilled with a bit that matches the diameter of the guide pin, serve as pilots.

and clamp the jig so the bottom of the V is aligned with the bit's center. When equally spaced holes are needed, mark the surface of the work with diameters that are the required number of degrees apart. Rotate the work so the diameters line up with the bit. Each layout line tells the location of a hole.

SURFACE HOLES

Radial surface holes can be located during layout using a compass to mark the circle and the spacing of the holes. However, it's more convenient to use a V-jig or pivot guides.

Clamp the V-jig to the table so it will be centered with the bit and so the work will be positioned for the hole's correct radial distance (FIG. 5-87). Rotate the work so a hole can be drilled on each radial line that you have marked. If the holes must go through the stock, mount the V-jig on a backup.

Pivot guides are jigs designed so the work can rotate on a central point. The distance from the pivot to the bit determines the radius of the circle on which the holes are drilled (FIG. 5-88). The work is impaled on the pivot, which can be a nail, and turned the required number of degrees for each hole.

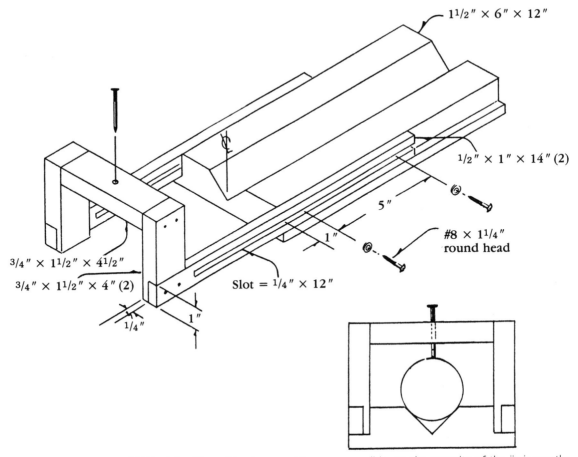

$1^{1/2}'' \times 6'' \times 12''$

$1/2'' \times 1'' \times 14''$ (2)

$5''$

$1''$

#8 × $1^{1/4}''$
round head

$3/4'' \times 1^{1/2}'' \times 4^{1/2}''$

$3/4'' \times 1^{1/2}'' \times 4''$ (2)

Slot = $^{1/4}'' \times 12''$

$1/4''$

$1''$

$1''$

5-82 A more advanced V-block jig. The guide pin assembly moves on slides so the capacity of the jig is greatly increased.

Making a simple jig that you can use anytime is just a matter of setting a length of dowel in a platform that can be clamped to the table (FIG. 5-89). The work must be center drilled to suit the dowel, but if you don't want a hole through the work, use a short dowel and drill a blind hole in the work. This also applies if a nail is used as a pivot point. Cut the nail so it projects above the platform, less than the thickness of the stock.

Figure 5-90 offers a plan for a more elaborate jig that has a slide support for the pivot. Thus, various radial distances can be established without having to relocate the jig. Note that, because the slide is equipped with a threaded sleeve, various pivot point designs are suggested. They can be long to pass through the stock, or short so the stock can just be impaled on them for rotating. The slide can be turned end-for-end so there is considerable flexibility here in terms of work size and the radial distance of holes.

For the surface hole jigs discussed so far, it's necessary to do preliminary lay-out to mark the spacing of the holes. The advanced jig shown in FIG. 5-91 incorpo-

5-83 The adjustable V-block jig in use. Guide pins do more than gauge hole spacing. They also keep the work from rotating and so assure that all holes will be on one centerline.

5-84 Using a V-block to drill spoke holes in a small wheel. The wheel is clamped to a backup that is tack-nailed to the block.

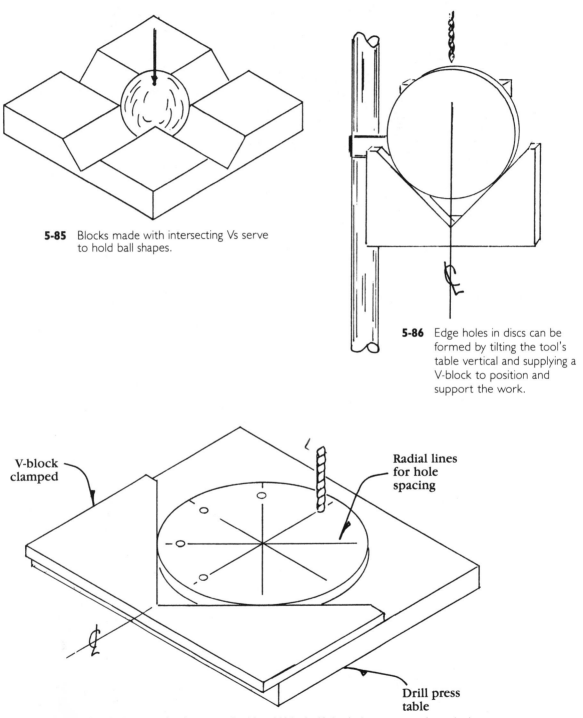

5-85 Blocks made with intersecting Vs serve to hold ball shapes.

5-86 Edge holes in discs can be formed by tilting the tool's table vertical and supplying a V-block to position and support the work.

V-block clamped

Radial lines for hole spacing

Drill press table

5-87 Radial surface holes can also be gauged with a V-block. If the holes must go through the stock, mount the V-block on a backup.

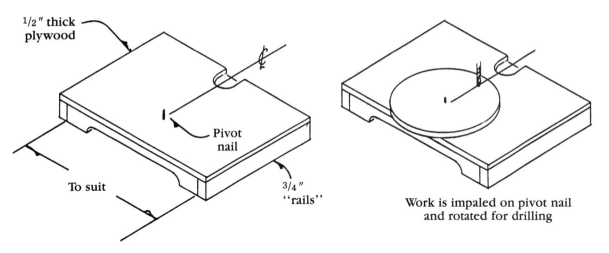

1/2" thick
plywood

Pivot
nail

To suit

3/4"
"rails"

Work is impaled on pivot nail
and rotated for drilling

5-88 A pivot jig can be just a platform with a nail through it on which the workpiece can be centrally mounted.

5-89 An elementary, permanent pivot jig can be an auxiliary table with a dowel glued in place. The jig is clamped to provide the required radial distance for the holes.

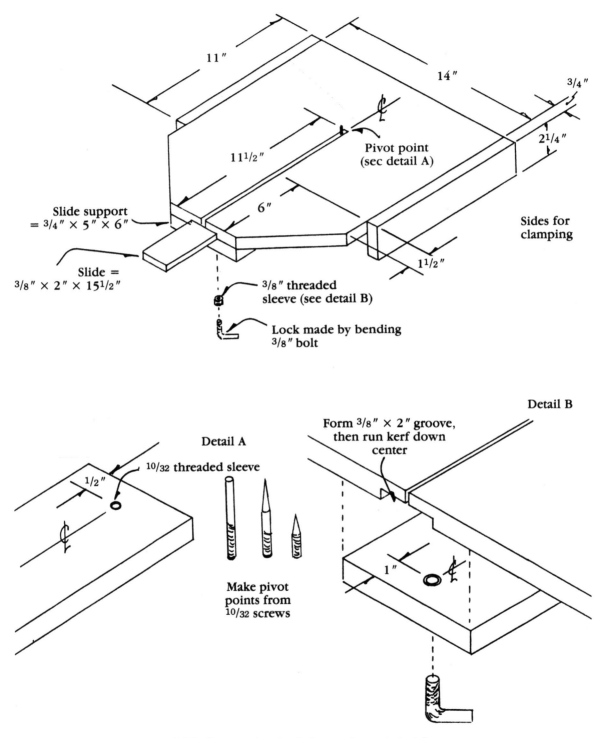

11"

14"

3/4"

11 1/2"

Pivot point
(sec detail A)

2 1/4"

Slide support
= 3/4" × 5" × 6"

6"

Sides for
clamping

Slide =
3/8" × 2" × 15 1/2"

1 1/2"

3/8" threaded
sleeve (see detail B)

Lock made by bending
3/8" bolt

Detail B

Form 3/8" × 2" groove,
then run kerf down
center

Detail A

10/32 threaded sleeve

1/2"

1"

Make pivot
points from
10/32 screws

5-90   Construction details for an advanced pivot jig.

5-91 The indexing jig negates the need to do layout for hole locations. After the first hole, the pin is pulled and then returned to engage another hole in the indexing plate.

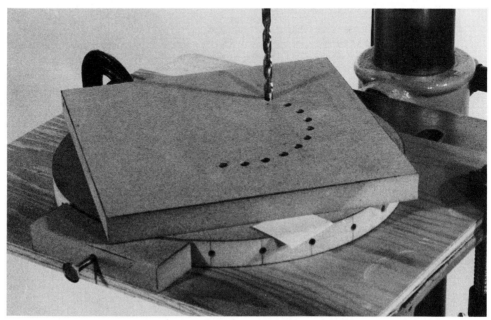

5-92 Use the indexing jig to drill square stock as well as discs. Use double-faced tape so the work and the indexing plate will rotate together.

rates an indexing device so that hole spacing will be automatic and accurate. Spacing of the edge holes in the indexing plate is variable, but about 10 degrees seems to be right for general use. You can drill fewer holes, or more, but the jig's accuracy depends entirely on how carefully you do that part of the job. There is no reason why the jig can't be used to drill square pieces as well as circular ones (FIG. 5-92). Notice that in both photos depicting the indexing jig in use, the workpiece is held to the indexing plate with pieces of double-faced tape. Figure 5-93 shows how the jig is made.

DRILLING HOLES FOR WOOD SCREWS

In order for wood screws to drive without requiring excessive torque, and to hold with maximum strength, they need two different size holes on a common, vertical centerline. Other factors to consider are whether the screw will be set flush with the surface of the stock, which involves countersinking, or whether a counterbore is required so the screw can be hidden with a plug or a decorative button (FIG. 5-94).

Because screws are generally available in many sizes and lengths (TABLE 5-2), it's necessary to size shank and lead holes to suit the screw. Follow the information provided in TABLE 5-3 when drilling body and lead holes for wood screws.

A common procedure when working with conventional drill bits is to first form the lead hole, which then serves as a guide for the body hole. The reverse procedure is also acceptable; the choice is a personal one. The body hole should

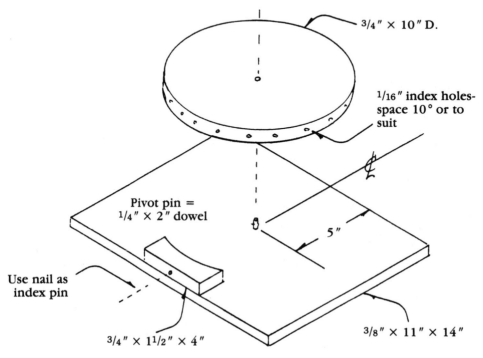

5-93 Construction details for indexing jig.

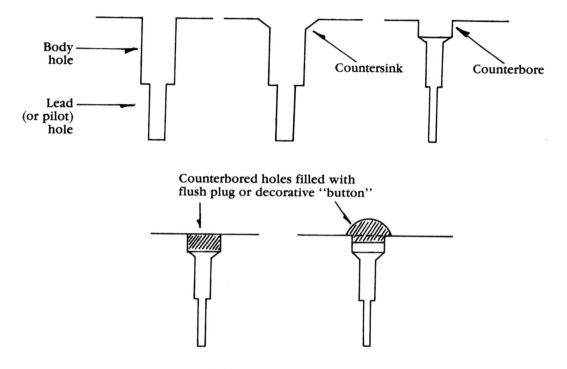

Body hole

Lead (or pilot) hole

Countersink

Counterbore

Counterbored holes filled with flush plug or decorative "button"

5-94 Nomenclature of screw holes.

Table 5-2 Sizes of Common Wood Screws

Number of Screw	Body Size Decimal	Lengths Available (Inches)
0	.060	1/4
1	.073	1/4
2	.086	1/4, 3/8, 1/2
3	.099	1/4, 1/2, 3/8, 1/4
4	.112	3/8, 1/2, 5/8, 3/4
5	.125	3/8, 1/2, 5/8, 3/4
6	.138	3/8, 1/2, 5/8, 3/4, 7/8, 1, 11/2
7	.151	3/8, 1/2, 5/8, 3/4, 7/8, 1, 11/4, 11/2
8	.164	1/2, 5/8, 3/4, 7/8, 1, 11/4, 11/2, 13/4, 2
9	.177	5/8, 3/4, 7/8, 1, 11/4, 11/2, 13/4, 2, 21/4
10	.190	5/8, 3/4, 7/8, 1, 11/4, 11/2, 13/4, 2, 21/4
11	.203	3/4, 7/8, 1, 11/4, 11/2, 13/4, 2, 21/4
12	.220	7/8, 1, 11/4, 11/2, 13/4, 2, 21/4, 21/2
14	.242	1, 11/4, 11/2, 13/4, 2, 21/4, 21/2, 23/4
16	.268	11/4, 11/2, 13/4, 2, 21/4, 21/2, 23/4, 3
18	.294	11/2, 13/4, 2, 21/4, 21/2, 23/4, 3, 31/2, 4
20	.320	13/4, 2, 21/4, 21/2, 23/4, 3, 31/2, 4
24	.372	31/2, 4

Table 5-3 Drill Sizes for Wood Screws

SIZE OF SCREW	BODY HOLE		LEAD HOLE HARDWOOD		SOFTWOOD	
	Fractional Size	Number or Letter	Fractional Size	Number or Letter	Fractional Size	Number or Letter
0	1/16	52	1/32	70		
1	5/64	47	1/32	66	1/32	71
2	3/32	42	3/64	56	1/32	65
3	7/64	37	1/16	54	3/64	58
4	7/64	32	1/16	52	3/64	55
5	1/8	30	5/64	49	1/16	53
6	9/64	27	5/64	47	1/16	52
7	5/32	22	3/32	44	1/16	51
8	11/64	18	3/32	40	5/64	48
9	3/16	14	7/64	37	5/64	45
10	3/16	10	7/64	33	3/32	43
11	13/64	4	1/8	31	3/32	40
12	7/32	2	1/8	30	7/64	38
14	1/4	D	9/64	25	7/64	32
16	17/64	I	5/32	18	9/64	29
18	19/64	N	3/16	13	9/64	26
20	21/64	P	13/64	4	11/64	19
24	3/8	V	7/32	1	3/16	15

approximate the diameter of the screw's shank, while the lead hole should be about 70 percent of the screw size in soft wood, 90 percent in hardwood.

Drilling holes for screws will be easier and more efficient if you take advantage of readily available bits that will provide all the configurations in one step. Various concepts are offered so there is no problem in choosing a design that suits the work you are doing (FIG. 5-95).

Among the products are those made by Stanley Tools with such names as "Screw Mate" that forms a countersink in addition to body and lead holes, and "Screw Sink" that automatically forms a counterbore. Another design that is becoming more popular works with a tapered bit, the idea being that the taper is more suitable for the thread portion of the screw (FIG. 5-96). The drill bits can be purchased alone or with adjustable sleeves that will add a countersink or a counterbore.

The usual size of the screw hole tools run from a 3/4-inch No. 6 screw, up to a 2-inch, No. 12 screw. They can be purchased individually or in sets. Sets are preferable because no single cutter will do for all the screw sizes.

With some of the tools the maximum, correct depth you can drill will be controlled by an integral collar. On others, an adjustable stop of some particular design will be supplied with the cutters or will be available as an accessory (FIG. 5-97).

Countersinking

A countersink is required when flat-head screws are driven flush with the work's surface and when the screw has an oval head (FIG. 5-98). When the countersink is

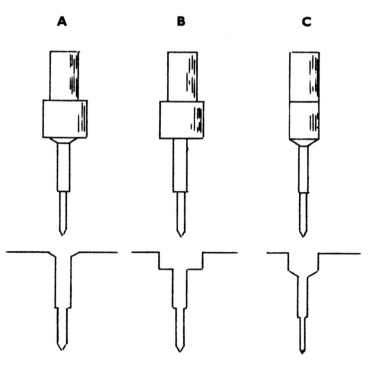

5-95 Cutters like these make it easy to drill wood screw holes accurately, and they do the entire job in one step.
(A) Forms countersink, body hole and lead hole
(B) Forms counterbore, body hole and lead hole
(C) Forms counterbore, countersink, body hole and lead hole

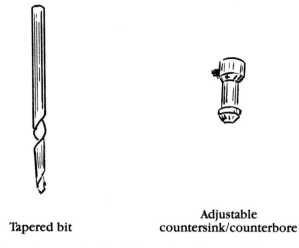

Tapered bit

**Adjustable
countersink/counterbore**

5-96 Tapered bits conform to the threaded portion of a screw. They can be purchased with adjustable countersink.

5-97   Stops limit how far a screw hole bit will penetrate. Often such a stop will be included with a set of bits.

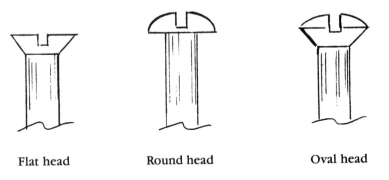

Flat head   **Round head**   **Oval head**

5-98   Typical screw heads. Flat and oval head screws call for countersinking.

not provided by a screw-hole bit, or the one needed is oversize, it is formed with a special cutter that cone shapes the top edge of the hole at an included angle (for wood screws) of 82 degrees (FIG. 5-99).

If the drilling chore will be executed just once, you can judge fairly accurately how deep to go. When many cuts are needed, use the stop rod to control quill extension so the cuts will have a uniform depth. Countersink to the full depth of the screwhead when working with hardwoods. On softwood, stay a bit shy because the screw will pull itself flush as you finish driving it.

5-99 Countersinking is done with a special cutter. Use the drill press stop rod to control depth when many similar countersinks are needed.

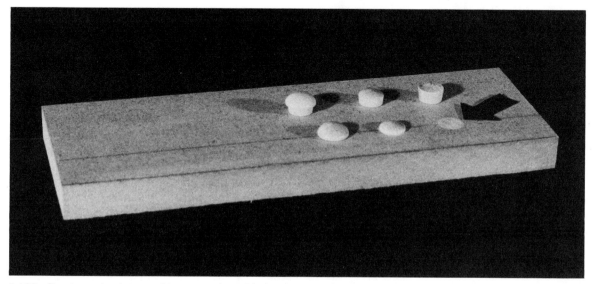

5-100 Ready-made plugs used in counterbored holes to conceal screws are made to be set flush, or to project as decorative details.

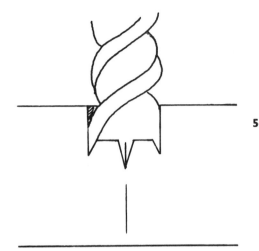

5-101   Conventional wood bits can be used for counterboring.

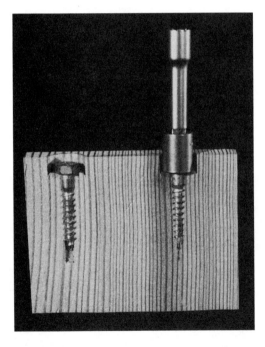

5-102   Counterbores for heavy-duty fasteners should be large enough to accommodate the socket that must be used to tighten the fastener.

It might not be necessary to provide a countersink when you are driving very small screws into softwood. Just providing a starting hole with an awl could be sufficient. You might be able to work the same way on some hardwoods. Judge for yourself by trying it with a screw or two.

Counterboring

A *counterbore* is a cavity that is sized to suit the diameter of the screwhead so the fastener can be driven below the surface of the wood and then concealed with a plug. The plugs can be custom made by using special cutters (see Chapter 7), or they can be purchased. Ready-made plugs include those that are set flush and others that project above the surface of the wood to provide a decorative detail (FIG. 5-100).

Counterboring can be done with tools made for the purpose or they can be part of the shape formed when using screw-hole bits. Actually, any counterbore can be formed by using conventional drilling tools (FIG. 5-101). When counterbor-

Table 5-4   Troubleshooting: Drilling

TROUBLE	CAN BE CAUSED BY	CHECK FOR OR DO
Work Splinters at Breakthrough	No backup	Use scrap block under the work
Holes not Perpendicular	Misalignment	Table must be 90° to centerline of spindle
Spindle Speed Seems Wrong	Belt slippage	Reset v-belts
Cutting Tools Overheat—Burn Marks on Work	Dull cutting tool	Sharpen or replace
	Waste chips accumulate	Retract cutting tool frequently to clear away waste
	Excessive speed	Use correct speed or stay as close as possible
	Feed/speed pressure	Feed only enough to keep cutting tool working
Bit Dulls Quickly or Cutting Edges Break	Feed is too slow	Feed so tool is cutting continuously too slow as bad as too fast
Work Tends to Twist	Poor procedure	Clamp work to table or to a fence
Bit Moves Off Center	Poor procedure	Mark hole location with punch or awl
	Drift	Drill pilot hole first, then enlarge— in stages if necessary
Cutting Tool Slips	Chuck not tightened	Always use the chuck key
Bit Binds	Work pinch	Always use a backup
	Feed pressure	Don't force the cut
	Hole too large	Drill pilot hole—increase in stages
Inaccurate Angular Hole	Poor procedure	Be sure work is at correct angle
	Edge of bit makes contact before its center does	Make initial contact very slowly— use correct angular support
Wrong Hole-Depth	Poor procedure	Recheck depth stop
	Overlooking point on bit	Allow for points on bits when setting up
Quill Does not Return Easily	Bad adjustment	Check owner's manual for adjustment procedure

ing is done to conceal a fastener, the diameter of the cavity must provide a snug fit for the plug.

There are occasions when it's necessary to provide an oversize cavity—for example, when you wish to conceal a heavy-duty fastener such as a lag screw. Then the counterbore must be large enough to provide room for the socket that will be used to drive the screw home (FIG. 5-102).

The depth of a counterbore should be a bit greater than the length of whatever will be used to fill it. This will provide some room for excess glue.

TABLE 5-4 offers possible solutions for most common basic drilling problems.

Chapter **6**

Drilling angular holes

Angularly drilled holes are those that enter a surface obliquely. This also includes holes formed in workpieces that are secured at an off-vertical angle, like those required in miter cuts when the components will be joined with dowels. Although the latter are not really angular holes because they enter the work surface squarely, they are included in this chapter.

There are, essentially, two categories of angularly drilled holes: *simple*, and *equal compound*. Visualize a project component, like a table leg or chair spindle, inserted in an off-vertical hole (FIG. 6-1). With a simple angle, the inserted piece will tilt in one direction, which will be obvious when you view it from one side. When the part is inserted in an equal compound-angle hole, its slant will be the same whether you view it from the front or the side. The centerline of the hole is always at a 45-degree angle. If the centerline is different—for example, at a 60-degree angle from an edge of the work—then you would have an *unequal* compound angle. The component still slants two ways but the tilt favors one direction.

PROBLEMS

A universal problem when drilling angular holes is getting the bit started accurately. This is not too difficult with a cutting tool such as a spade bit, because it has a long point that will contact the work before cutting begins (FIG. 6-2). Other hole formers such as brad point bits have short center points, so unless the angle of the hole is very slight, the spurs of the bit will make contact before the tool is set. This can cause the bit to wander from the hole's center. This also applies to bits like twist drills. Their cutting edges are sloped so they have a tendency to slide when pressed against a slanted surface.

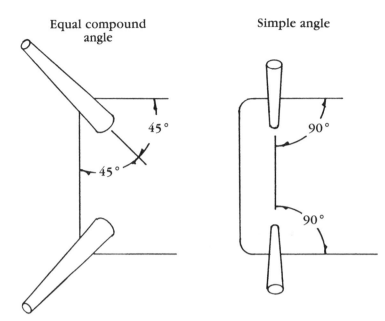

Equal compound angle

Simple angle

6-1 The direction in which a component will tilt depends on the slant direction of the hole. The amount it will tilt depends on the angle at which the hole is drilled.

6-2 Spade bits have long points that contact the work before cutting starts.

The solution is to be extremely careful when the bit first contacts the work-piece or, when possible, to provide a leveling block so the bit will enter a flat surface even though the work is in a slanted position. Forming "pocket holes," slanted cavities that are required when screws are used to, for example, connect a tabletop to rails, is a good way to introduce the practice (FIG. 6-3).

One method employs holding blocks that have been beveled to hold the work at the correct angle (FIG. 6-4). This works best when the surface of the front block is level with the point at which the bit will contact the work.

Another idea is to make a jig for permanent use (FIG. 6-5). Saw a slanted groove to receive the work into a base that can be clamped to the drill press table. Both the backup and the leveler can be permanent attachments. The hole in the leveler, which will be formed when the first hole is drilled, provides guidance and keeps the bit from wandering when other holes are formed.

The second problem has to do with the size of the component. It's difficult to place something like a tabletop on a drill press table so it can be drilled accurately. An easy solution is to do the drilling in blocks that can be attached to the project to receive the part that will be inserted. Even though the holes in the blocks are drilled at a simple angle, the way you position them will determine the direction in which the component slants (FIG. 6-6).

TILTING TABLE

When the drill press table can be tilted, the drilling angle can be established by using a T-bevel (FIG. 6-7). Because tables tilt on a plane that is perpendicular to the column, they aren't very convenient for drilling angular holes on long pieces of stock. The Shopsmith multipurpose machine, in its vertical drill press mode, is an

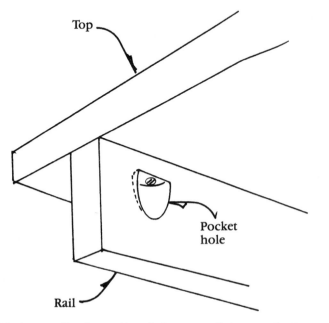

Top

Pocket
hole

Rail

6-3 Pocket holes are often formed in rails for screws that secure the top component.

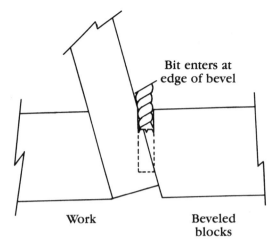

Bit enters at edge of bevel

Work

Beveled blocks

6-4 Leveling blocks are used so that the bit will have a flat surface to enter.

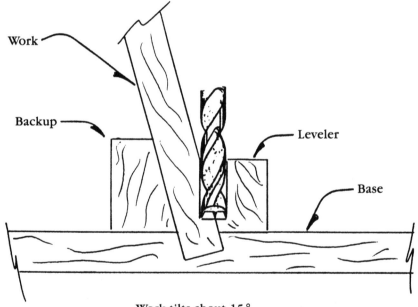

Work

Backup

Leveler

Base

Work tilts about 15°

6-5 A leveling jig like this one can be maintained as a permanent accessory. It is designed, of course, for a particular stock thickness.

exception. Because of the manner in which the table tilts, the length of the work is immaterial (FIG. 6-8). A positive factor is that the machine's rip fence can be used to support the work. This also assures that holes will be on a common centerline.

One technique that allows angular drilling of long pieces on a conventional table is demonstrated in FIG. 6-9. The front strip of wood serves to support the work, while a second strip under the work allows it to be tilted. The thickness of the height strip and its position determine the angle of the hole.

| Equal compound angle | Simple angle | Any other angle |

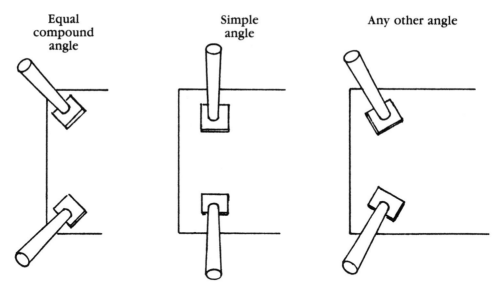

6-6 Angularly drilled blocks can be used to establish the slant direction of components.

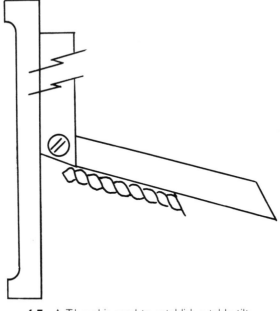

6-7 A T-bevel is used to establish a table-tilt.

A jig for angular drilling

The project detailed in FIG. 6-10 provides an auxiliary tilting table that can be secured with clamps to a regular table. The platform and the hinged panel, which is actually the worktable, are made of plywood—which isn't the best material for holding screws that are driven into its edges. For this reason, it's probably wise to

6-8 The Shopsmith table tilts in a way that does not limit the length of workpieces that can be drilled.

make the two components as shown in FIG. 6-11. The lumber strips will do a better job of holding the screws that are used to attach the piano hinge.

A strip of wood, tack-nailed to the table, will serve as a fence to support the work. When possible, it should be placed at the front edge of the jig (FIG. 6-12). If

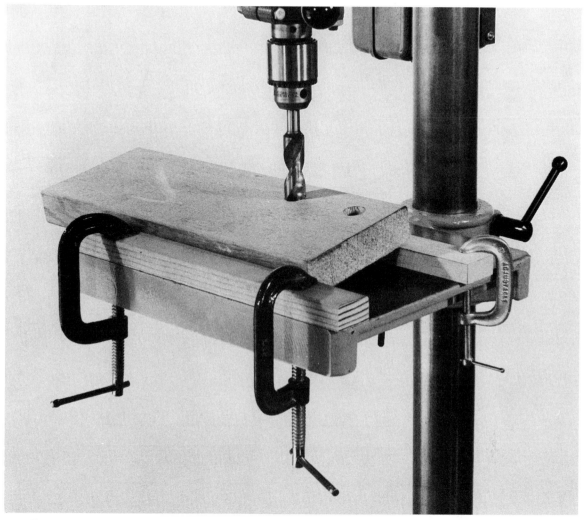

6-9 This improvised setup is a practical way to do angular drilling. The thickness and position of the height strip determine the slant angle of the hole.

the width of the work doesn't allow this, the fence can be placed at the back edge of the tilting table. In this case, you must hold the work so it doesn't slide out of position. Figure 6-13 shows how a V-block can be used with the jig to hold circular components for drilling radial, angular holes.

DRILLING INTO MITERS

Miter joints that are reinforced with dowels require holes that are perpendicular to the miter cut. One way to do this with a tilting table is to set the table at 45 degrees and to position the basic fence (or a strip of wood) so the work can be clamped in correct position for drilling (FIG. 6-14). A stop block will assure that the hole in all pieces will have the same location. Two holes are usually required in this type of

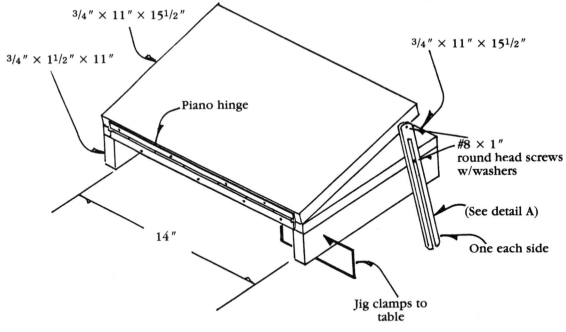

$3/4'' \times 11'' \times 15^1/2''$

$3/4'' \times 1^1/2'' \times 11''$

$3/4'' \times 11'' \times 15^1/2''$

Piano hinge

#8 × 1"
round head screws
w/washers

(See detail A)

One each side

14"

Jig clamps to
table

Detail A

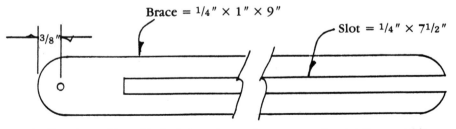

Brace = $1/4'' \times 1'' \times 9''$

Slot = $1/4'' \times 7^1/2''$

3/8"

6-10   Plans for a tilting table jig that can be used on a conventional drill press table.

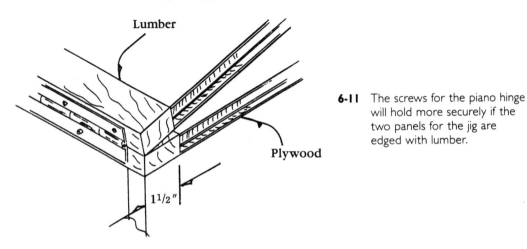

Lumber

Plywood

$1^1/2''$

6-11   The screws for the piano hinge
will hold more securely if the
two panels for the jig are
edged with lumber.

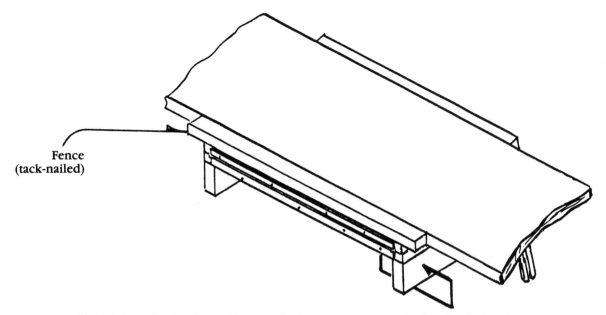

**Fence
(tack-nailed)**

6-12   A fence for the tilting table jig can be just a strip of wood that is tack-nailed in place.

6-13   Using a V-block with the tilting table jig to drill angular, radial holes in a circular component.

6-14 A setup like this can be used to drill into miter cuts if the table can be tilted 45 degrees.

joint. Drill the first hole in all the parts without disturbing the setup. Then relocate the stop block so the pieces can be positioned for the second hole (FIG. 6-15).

A jig for drilling into miters

Drilling dowel holes in miter cuts can be done accurately and more conveniently by making a jig that can be maintained as a permanent accessory. The jig, shown in FIG. 6-16, can handle work of various widths and in lengths that are not greater than the distance from the spindle to the floor. The position of the jig determines where the holes will be, while a clamped fence against which the jig rests establishes edge distance.

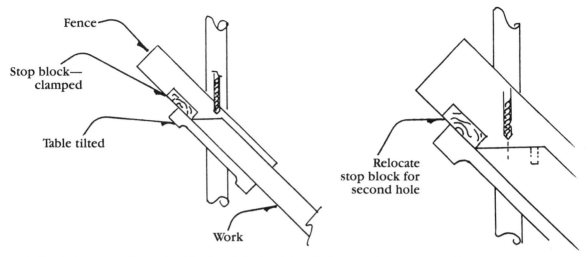

Fence

Stop block— clamped

Table tilted

Work

Relocate stop block for second hole

6-15 Two holes are usually required for doweled miter joints. Drill the first hole in all parts. Then change the position of the stop so parts will be set correctly for the second hole.

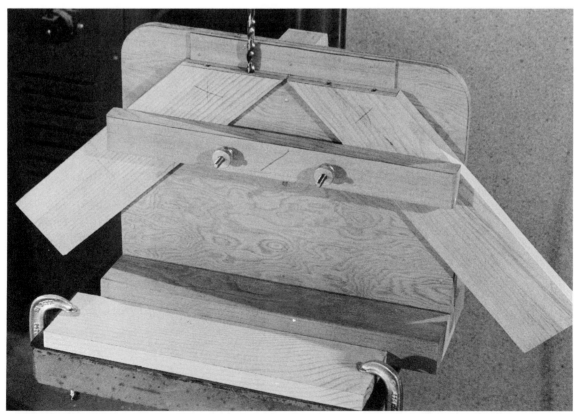

6-16 This special jig is fine for drilling dowel holes in miters because it handles both left- and right-hand cuts.

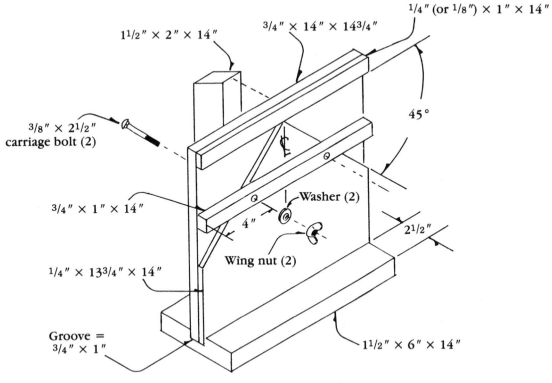

6-17 Plans for the special jig.

The jig is designed to handle left- or right-hand miter cuts, so there can be some leeway in whether the holes are exactly in the center of the stock's thickness. Make a mark on all pieces to identify the "good" surface. When you position the pieces for drilling, be sure the marked surface faces either forward or back. It won't matter then if the holes are a bit off, so long as you do the assembling with the marked sides of all the parts either up or down.

Construction details for the jig are shown in FIG. 6-17.

Chapter **7**

Working with plug cutters

Plug cutters are practical drill press tools that form precise cylinders of various diameters in hard or soft wood, regardless of whether cutting is done with or across the grain. The example displayed in FIG. 7-1 is literally a plug cutter, because it is designed for short cylinders that are mainly used in counterbored holes to conceal screws. The example in FIG. 7-2 is also called a plug cutter, but because it can produce cylinders up to about 2 inches long, it's very handy for making dowels that can be used in joints.

The tools are gripped in the chuck and used much like drill bits. They are confined in the ring they form to produce the cylinder, so it's wise to retract more frequently than you do with a conventional drill bit, to prevent overheating of cutter and work. Speed should be adjusted to suit the size of the cutter and the density of the material. Starting points are offered in TABLE 7-1, but you can make adjustments one way or the other if the tool doesn't cut steadily and cleanly with reasonable feed pressure. Having to bear down too hard will probably indicate that the speed is wrong or that the cutter is dull.

FORMING PLUGS

Anyone who has cut plugs from commercial dowels will appreciate the advantages of custom-made units. It's often difficult to find dowels that will match the wood species used in various projects. The diameters of those available in local hardware stores are seldom precise and, because the grain runs lengthwise, using one to conceal a fastener will appear obvious when the project is stained.

Plug cutters are precise and can cut into surfaces so that the plug will have a cross-grain pattern that flows with adjacent surfaces and—important for finishing—a similar surface density. Plugs can be produced from any wood species so

7-1 The plugs produced with this cutter have one end that is slightly rounded. The shaped end can project above the stock as a decorative detail, or, if the plug is inverted, will provide some room in the hole for excess glue.

7-2 This cutter can form short plugs or joint dowels up to about 2 inches long.

Table 7-1
Suggested Spindle Speeds (RPM)
When Using Plug Cutters

Size	Softwood	Hardwood
3/8 "	2400	1250
1/2 "	1250	1250
5/8 "	1250	1250
3/4 "	1250	700
1 "	700	700

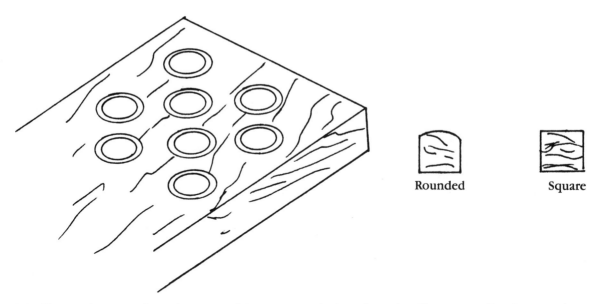

Rounded   Square

7-3 Plugs can be cut so the grain runs parallel to or across the long dimension. They are usually cut cross-grain so they will conform with the grain flow of the work.

the problem of matching plug and work is eliminated. Also, depending on the cutter that is used, plugs can have flat ends or those with a slightly rounded contour at one end (FIG. 7-3).

Cuts can be made through any material thickness within the cutter's capacity, even through veneers when you need to patch a flaw or require a small disc for an inlay. Cutting through works, but it involves having to stop and retrieve each plug as it is cut. When many are needed, it's more convenient to use the method shown in FIG. 7-4. Cut into stock that is thicker than the length of the plug and then free the plugs with something like a screwdriver (FIG. 7-5). The broken end will not be uniform but that is not critical because it will not be visible. In fact, the irregularity will allow room for excess glue when the plug is installed.

Cut plugs that are used to conceal fasteners slightly longer than necessary so they can be sanded flush after they are installed. Also, on critical work plan the cuts very carefully so the grain pattern will flow as smoothly as possible (FIG. 7-6).

FORMING DOWELS

Here "dowel" will be used to identify the wooden cylinders that are used in joints. It is not a good idea to cut these from dowel rods that are available in 3- or 4-foot lengths because their diameters are seldom precise. One that is over- or undersize can be a nuisance at assembly time. Commercial dowels made specifically for joint use (often called "dowel pins") are available in the sizes shown in FIG. 7-7. These are precise in diameter and length and are treated to provide escape routes for excess glue (FIG. 7-8). Commercial dowels have their benefits, but by having a set of cutters on hand, you can save money and can be prepared for any

7-4 It's best to make stopped cuts when many pieces are needed.

chore without having to run to a store or wait for a mail delivery. You are also relieved of having to supply storage for an assortment of dowels.

One of the popular products for forming dowels is shown in FIG. 7-9. This precision tool is available for dowels up to about 2 inches long and in diameters of ³/₈, ¹/₂, ⁵/₈, ³/₄, and 1 inch. It can be used to cut through stock but it's better to follow the method that was suggested for making stopped cuts (FIG. 7-10). In this case, it's better to free the dowels by making a saw cut, either on a power tool like a table saw or with a handsaw (FIG. 7-11).

Usually it's better to cut into end grain so the dowel will have a strong cross section (FIG. 7-12). The dowels are finished by slightly chamfering both ends and by providing a means for excess glue to escape from the hole. You can accomplish that by using a small carving chisel to form a longitudinal groove or simply by using a pair of pliers with serrated jaws to indent the dowel in various places (FIG. 7-13).

A good joint calls for installing two dowels (FIG. 7-14). This doubles the strength of the connection and provides an anti-twist factor. Incidentally, there is no reason why the dowel former can't be used to produce short plugs (FIG. 7-15).

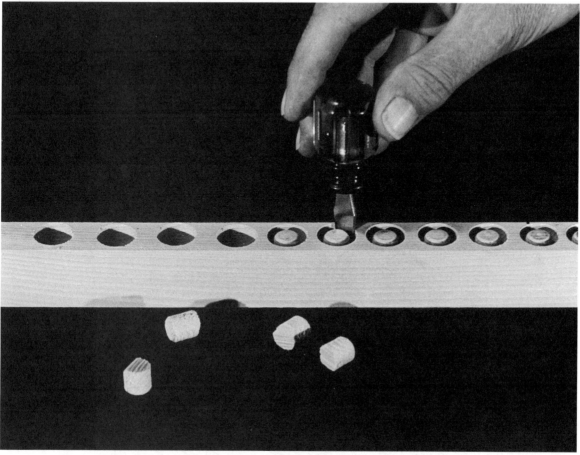

7-5 The captured plugs are freed by prying them off with a screwdriver.

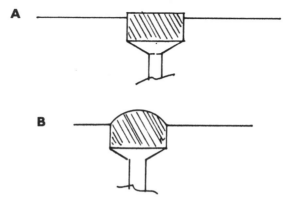

7-6 (A) Plugs that project may be sanded flush after glue dries. (B) Rounded plugs provide decorative detail.

Diameter	Length
1/4"	1½"
3/8"	1½", 1¾", 2"
7/16"	2"
1/2"	

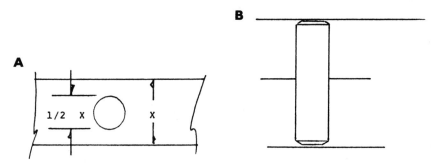

7-7 The chart shows available sizes of commercial dowel pegs. The maximum peg diameter should be half the stock thickness (A) and ¹/₃₂ to ¹/₁₆ inch shorter than total hole depth (B).

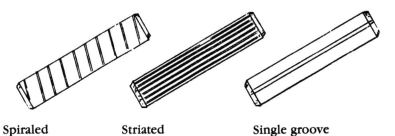

Spiraled   Striated   Single groove

7-8 Commercial dowels are treated in one of these ways. The indents allow glue to spread and provide room for excess amounts.

FORMING INTEGRAL TENONS

A unique application for plug cutters is forming integral, round tenons on square or round stock. The basic technique is shown in FIG. 7-16. Kerfs are sawn into four sides of the component to a depth that will meet the circumference of the tenon. On square stock, the kerfs can be formed on a table saw. If you have the expertise, you can do the kerfing on round stock by using a scroll saw or bandsaw. In either case, of course, the kerfing can be done with a handsaw.

7-9 This cutter will produce short plugs, but its major ability is producing dowels up to 2 inches long.

7-10 Dowels may also be formed by making stopped cuts, especially when many are needed.

The final step is to use a holding jig for the work and to position it in the drill press so the plug cutter can be used (FIG. 7-17). Both phases of the job call for careful work. If the kerfing is done accurately and the work is positioned precisely

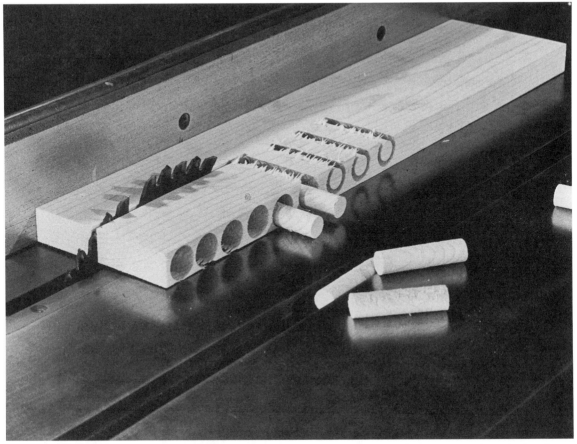

7-11 The dowels are set free with a saw cut. This can also be done on other power tools, or by hand.

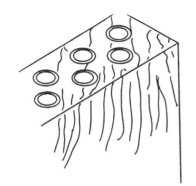

7-12 Joint dowels should always be cut into end grain so they will have a strong cross section.

under the plug cutter, most of the waste will fall away. The result will be a neat, integral tenon. The tenons do not have to be centered. You can place them where you wish simply by doing the initial kerfing accordingly.

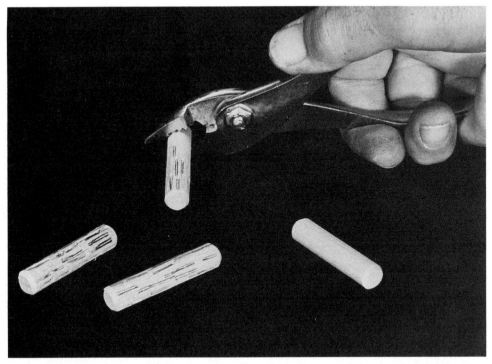

7-13 A simple way to provide for excess glue. Form indents by squeezing dowels with pliers that have serrated jaws.

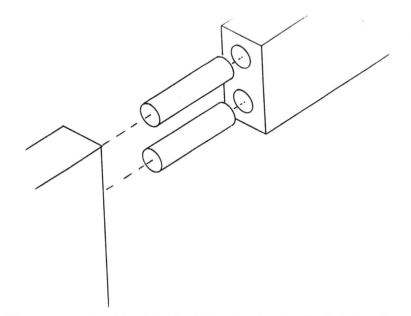

7-14 Always use two dowels in a joint. In addition to extra strength, the twins will oppose any twisting force.

7-15 The dowel former can also be used to form short plugs.

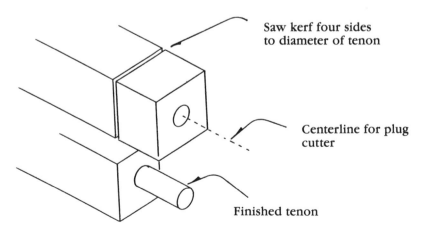

7-16 The procedure for forming integral tenons starts with kerfs that will meet the diameter of the tenon.

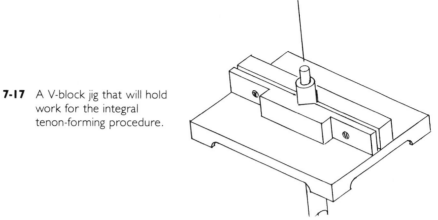

7-17 A V-block jig that will hold work for the integral tenon-forming procedure.

7-18 Integral tenon jig No. 2. You can skip the initial kerfing step when forming the tenons on short rounds.

Two jigs that you can make for this type of work are shown in FIGS. 7-18 and 7-19. Both work like split clamps to secure workpieces. The major difference between the two is that one is secured on a base and is drilled so the bolts that squeeze the clamp jaws together can be placed near the work. When making this version, be sure that only one jaw is attached to the base. The other must be free

7-19 Integral tenon jig No. 3. If you adopt this design be sure to secure only one half of the split clamp to the base.

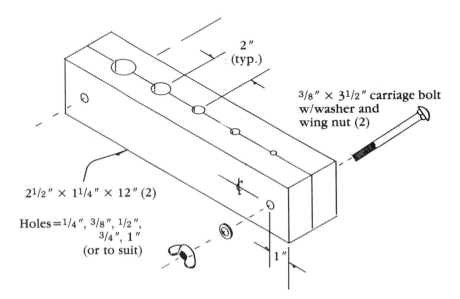

2" (typ.)

3/8" × 3½" carriage bolt w/washer and wing nut (2)

2½" × 1¼" × 12" (2)

Holes = 1/4", 3/8", 1/2", 3/4", 1" (or to suit)

1"

7-20 Plans for the No. 2 integral tenoning jig.

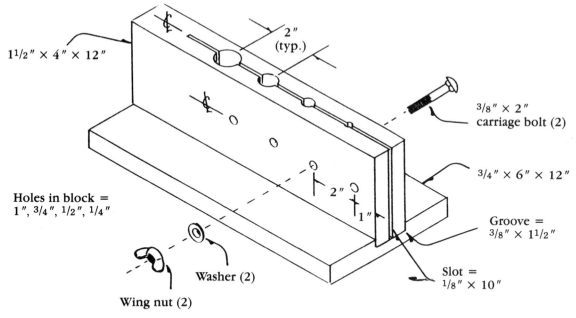

1½" × 4" × 12"

2" (typ.)

³/8" × 2" carriage bolt (2)

³/4" × 6" × 12"

Groove = ³/8" × 1½"

Slot = ¹/8" × 10"

Holes in block = 1", 3/4", 1/2", 1/4"

2"

1"

Washer (2)

Wing nut (2)

7-21 Plans for integral tenoning jig No. 3.

7-22 The jigs can be used to grip cylinders for concentric drilling.

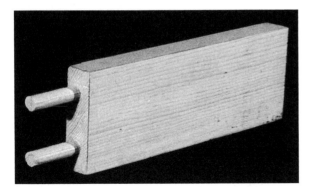

7-23 The basic technique can be used to form twin tenons.

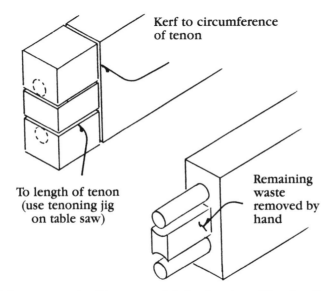

Kerf to circumference of tenon

To length of tenon (use tenoning jig on table saw)

Remaining waste removed by hand

7-24 Much of the waste stock will be removed during the plug cutting step, the remainder will have to be removed by hand.

to move when the lock bolts are tightened. A modification you might consider is to substitute matching V-cuts for the holes. Then the jig can be used to secure square stock as well as round (FIGS. 7-20 and 7-21).

Figure 7-22 demonstrates another application for the jigs: gripping work for concentric drilling.

The integral tenon technique can be used to produce twin tenons (FIG. 7-23). The starting procedure is the same, but while a good deal of the waste might fall away during the plug cutting step, some will remain that must be removed by hand (FIG. 7-24).

Chapter **8**

Mortising for classic joints

The mortise-and-tenon joint is a standard wood connection used in quality furniture. The design opposes radial stress and there is a substantial gluing surface to keep the parts together. The mortise is a rectangular or square cavity that can be blind or pass through the work; the tenon is an integral projection on the mating component that fits the mortise precisely (FIG. 8-1). There are variations (FIG. 8-2), but in all cases the components are essentially similar.

The tenons are formed by sawing, usually on a table or radial arm saw, but the mortises are drill press applications unless you tackle the difficult and time-consuming job of producing them with hand tools. It's always good practice to first form the mortise, then shape the tenon to fit.

TOOLS

Drill press mortising requires special accessories that are usually offered in kit form. The parts include a casting that attaches to the quill in some manner, a fence, and a hold-down to keep work secure (FIG. 8-3). How the parts are combined and situated on the machine can vary from tool to tool. On a Shopsmith, for example, the saw table rip fence is used, so necessary additional attachments include the casting and the hold-down (FIG. 8-4). In all cases, the standard three-jaw chuck is used to grip the cutting tools. The way in which the tools are situated in the casting varies; check the owner's manual for specific instructions.

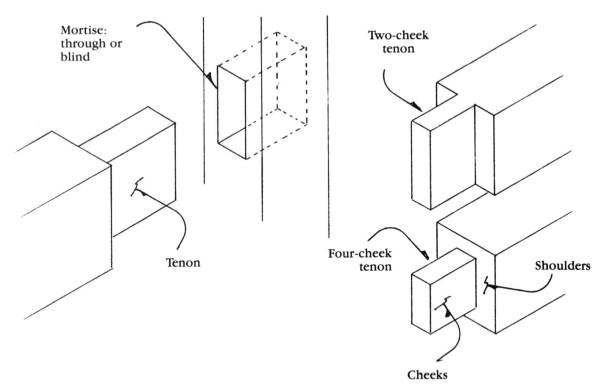

8-1 The mortise-and-tenon joint. Forming the mortise is a drill press application; tenons are shaped by sawing.

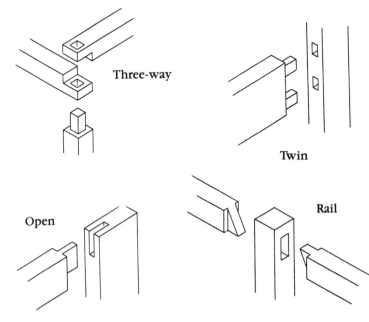

8-2 Variations of the mortise-and-tenon joint.

8-3 The mortising attachments for a drill press consist of a casting that attaches to the quill, a fence, and hold-downs. The cutting tools are purchased separately.

THE CUTTERS

The cutters are a two-part team that consist of a hollow, square chisel and a deeply fluted bit that rotates inside it (FIG. 8-5). The bit, working much like any drilling tool, removes the bulk of the waste, while the chisel cleans out the corners. The result is a square hole (FIG. 8-6). A rectangular cavity is achieved simply by making additional, overlapping cuts.

Chisels and bits are purchased as matching units. They may be purchased in individual teams or in sets that generally include sizes of $1/4$, $5/16$, $3/8$, and $1/2$ inch. The size of the chisel does not limit the cavity it can form. Just as overlapping cuts can form a rectangular shape, so can they produce a cavity that is wider than the chisel itself.

SETTING UP

There are several important factors to consider, regardless of the type of equipment you will use. The bit must be secured to allow some open space between its

8-4 On a Shopsmith, the machine's table saw fence is part of the arrangement. The hold-down locks in a hole that is in the fence.

cutting spurs and the sharp edges of the chisel (FIG. 8-7). The clearance, which should be at least 1/32 inch but not over 1/16 inch, is necessary to minimize heat build-up. Excessive friction between bit and chisel can do damage to the tools and the work. On the other hand, too much clearance can cause jagged edges where the bit enters the work, and result in large waste chips that can clog inside the chisel.

8-5 The cutting tools consist of a hollow, square chisel and a bit with deep flutes for easy waste removal. The bit rotates inside the chisel.

8-6 Bit and chisel work as a team to produce a square hole.

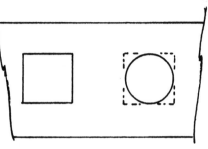

8-7 Clearance is needed to minimize friction between bit and chisel. It should be at least $1/32$ inch, never more than $1/16$ inch.

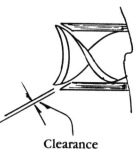

Clearance adjustment

The chisel must be organized so the angle between its side and the fence is 90 degrees. Check this with a square (FIG. 8-8), or by using the system that is shown in FIG. 8-9. Abut the chisel against a flat piece of wood that is clamped to the fence, or against the fence itself. You will know when this adjustment is not correct: The mortise you form will have serrated edges if done incorrectly (FIG. 8-10).

Also, be sure the table is in true horizontal position. The angle between the table's surface and the side of the chisel must be 90 degrees. If it isn't, you will produce mortises with slanted ends (FIG. 8-11).

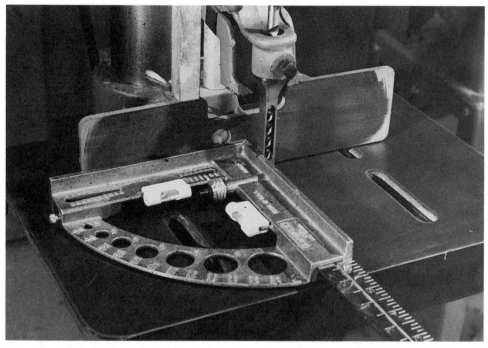

8-8 The chisel must be locked in the casting so that the angle between its side and the fence is 90 degrees.

8-9 Squaring the chisel can be done by abutting it to a flat board that is held tightly against the fence.

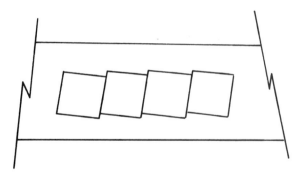

8-10 The mortise will have serrated edges if the alignment procedure isn't done carefully.

8-11 The mortise will have slanted ends if the angle between bit and table surface is not 90 degrees.

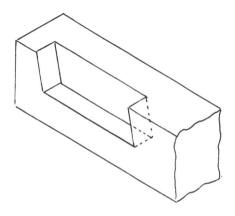

SPEEDS AND FEED

What applies to simple drilling also applies to mortising. Generally, the larger the chisel, the slower the speed that should be used, especially when working on hardwood. TABLE 8-1 suggests speeds that apply to hard or soft wood. Accept them as a starting point because there are many variations in wood species, even in different pieces of lumber that came from the same tree. Another factor is the

Table 8-1
Suggested Spindle Speeds (RPM)
When Mortising

SIZE OF MORTISE	SPEED (RPM)	
	Softwood	Hardwood
1/4"	2400	1250
5/16"	2400	1250
3/8"	2400	1250
1/2"	1250	1250

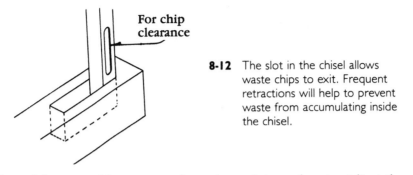

For chip clearance

8-12 The slot in the chisel allows waste chips to exit. Frequent retractions will help to prevent waste from accumulating inside the chisel.

direction of the cut: with or across the grain, or into end grain. Adjust the speed one way or the other until the tool is cutting efficiently.

The chisel does its job by being forced down into the work, so mortising requires a little more feed pressure than simple drilling, but always use the least amount that does the job. Having to muscle the cutters into the work is a sign that the speed is wrong or that chisel or bit (or both) are dull.

You have organized correctly and are working with adequate speed and feed pressure when waste chips travel easily up the flutes of the bit and emerge nicely from the escape slot in the chisel (FIG. 8-12). Frequent retractions will help things go smoothly.

1. **Make end cuts first**

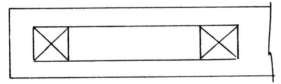

2. **Clean between with overlapping cuts**

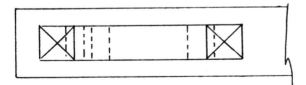

3. **On a wide mortise do not leave narrow shoulder. Use smaller chisel so cuts can overlap**

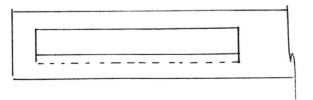

8-13 The basic cut sequence for forming mortises.

CUTTING THE MORTISE

A successful mortise depends on a series of cuts that are done in a particular sequence (FIG. 8-13). First, mark the work to indicate the length of the mortise. Then place the work so it will be firmly down on the table and against the fence, and adequately held by whatever hold-downs are available. The main hold-down is usually U-shaped and bears on the surface of the work to prevent it from lifting when the chisel is retracted.

The first consideration is preventing the chisel from slanting toward a cavity that has already been formed. Thus, it's wise to start by making end cuts first (FIG. 8-14). The waste is removed with additional cuts that overlap, ideally, about three quarters of the chisel's width (FIG. 8-15). The purpose of overlapping is to keep the chisel cutting on a perpendicular line.

When you are forming blind mortises, use the machine's stop rod to control cut depth. There will always be some splintering when the chisel emerges on mortises that go through the workpiece. The flaw can be minimized by using a backup, but for perfect work it's better to cut the through mortise on stock that is a bit thicker than the project requires. When mortising is finished, you can make a cut on another power tool like a jointer or table saw to shave off the imperfection.

Mortises that are wider than the size of a chisel are produced by making repeat, overlapping cuts on parallel lines (FIG. 8-16). Don't leave a narrow shoulder for the second line of cuts because it's likely that the chisel will move away from where it should be. This will cause the mortise to have a tapered side. Work with a chisel that will allow enough overlapping of cuts. For example, use a 3/8-inch

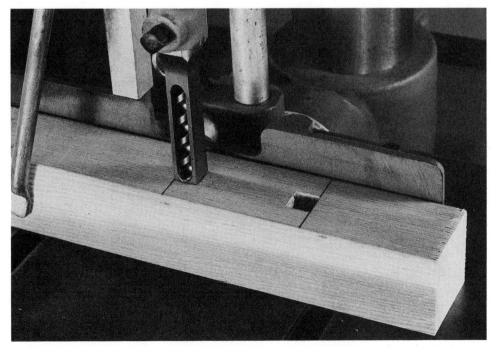

8-14 Always make the end-cuts first. (The hold-down was excluded here only to make the cuts more visible.)

8-15 Complete the mortise by using overlapping cuts to remove the waste. The purpose of the arm at the left, which is included with some mortising attachments, is to hold the work against the fence.

8-16 Wide mortises are formed by making parallel cuts. Select a chisel width that will not leave a narrow shoulder for the second line of cuts.

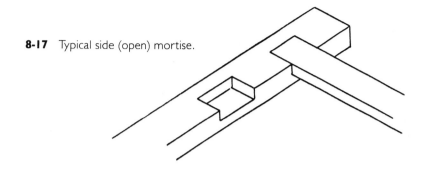

8-17 Typical side (open) mortise.

instead of a ¹/₂-inch chisel if the mortise must be ⁵/₈ inches wide. When the same mortise is required on several pieces, make the first line of cuts on all parts before resetting the fence to provide guidance for the final cuts.

Side mortise

Side mortises are used when project components are joined as shown in FIG. 8-17. The cuts, whether they are blind or through, are made in routine fashion, except that a backup is used (FIG. 8-18). The backup and the work should be held snugly together and against the fence (FIG. 8-19). When a shallow side mortise is needed, it will be cleaner and more precise if you use a wide chisel, allowing the excess width to cut into the backing.

A

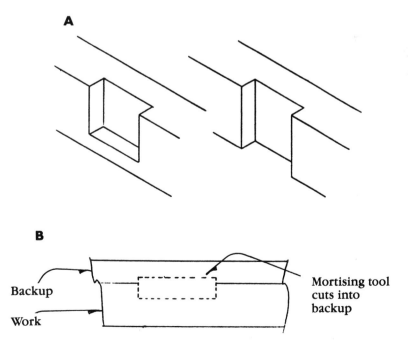

B

Backup

Work

Mortising tool cuts into backup

8-18 (A) Side mortises can be through or blind. (B) Use a backup between work and fence.

8-19 Backup in use.

Open-end mortise

Producing an open-end mortise is a routine operation. Start the cut at the open end, using most of the chisel's width, then clean out waste with overlapping cuts (FIG. 8-20). When a few pieces are needed, clamp a stop block to the fence to serve as a gauge for the length of the mortise.

LOCKING THE MORTISE

Even though mortise-and-tenon joints fit together tightly, they are often *keyed* or *wedged*, or treated in some manner for additional strength and to prevent separation caused by glue failure. Sometimes the added parts or *tusks* serve as a decorative detail; other times, the part is sanded flush or is simply hidden because of the joint design. This is the case when wedges are used in a blind mortise-and-tenon joint (FIGS. 8-21 through 8-24).

MORTISING ODD SHAPES

Any part that can be accurately positioned and securely held under the cutting tools can be mortised (FIG. 8-25). Many times, a clamped fence to which the work can be clamped serves adequately. Other times, the setup might have to be more creative to suit the job.

8-20 An open-end mortise doesn't require any special procedure. A stop block can be clamped to the fence to gauge the length of the cut—good practice when several parts are needed.

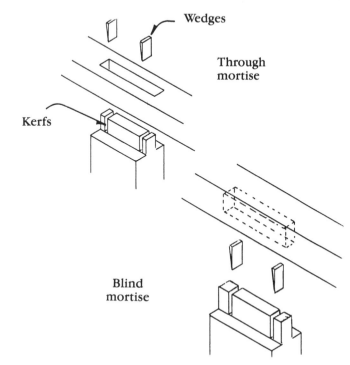

8-21 Wedges are often used to reinforce the joint. They force the tenon tightly against the sides of the mortise.

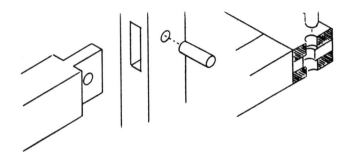

8-22 A dowel peg, used this way, will pull the tenon tightly against the mortised component.

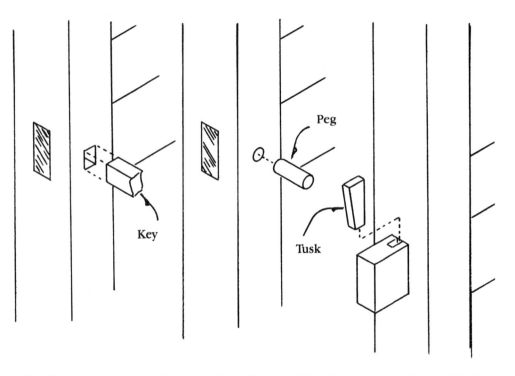

8-23 Pegs and keys are usually sanded flush. Tusks, which pull the tenoned piece tightly into place, are usually left exposed to provide a decorative detail.

The pivot jig concept that works for drilling radial holes on a circumference will serve just as well for forming mortises (FIG. 8-26). Notice that the mortising fence backs up the pivot jig's platform and that the hold-down performs its usual function.

Mortising can even be done concentrically into cylinders, if the work is adequately secured (FIG. 8-27). Mortising in rounds can go beyond this single application. Chapter 13 will include plans for a special jig that makes cylinder mortising a practical chore.

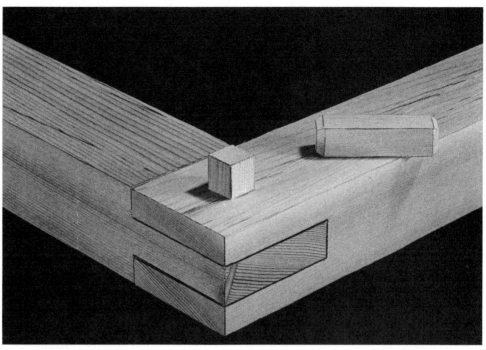

8-24 Mortising can be done for square pegs that reinforce other joint designs.

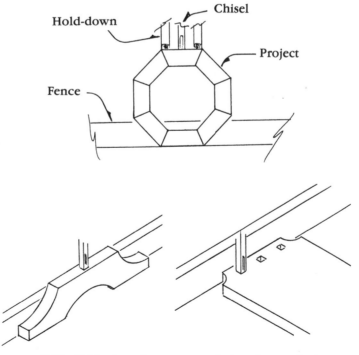

8-25 Oddly shaped components that can be mortised.

8-26   Using a pivot jig to form mortises on a circumference.

8-27   The modified handscrew acts as a vise to grip cylinders for concentric mortising.

8-28 Mortising can provide square corners for internal cutouts. Final cutting is done on a scroll saw or with a portable saber saw.

8-29 Using the mortising system to form fingers. Components must be trimmed so top and bottom edges will be flush.

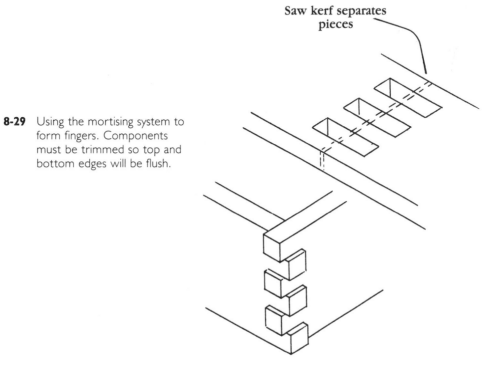

Saw kerf separates
pieces

Mortising can be utilized to make square cutouts that can be part of a design or that provide square corners for an interior opening in a project. The idea, as shown in FIG. 8-28, is to do the mortising first. The openings serve as insertion holes for scroll saw or saber saw blades, the tools that will be used to finish the job.

MORTISING FINGERS

An interesting variation of the mortising technique is forming shapes that can intertwine just like a fingerlap joint. The work starts by forming mortises that are more than twice as long as the stock's thickness. The stock is then separated on the centerline of the mortises. The parts can be assembled (FIG. 8-29). Some trimming will be needed so top and bottom edges of the assembly will be flush.

The idea can also be adapted to produce the connection shown in FIG. 8-30. In this case, the fingers pass through openings that are straightforward mortises. The ends of the fingers can project, as shown, or they can be sanded flush.

DRILLED MORTISES

Mortises that don't have to apologize to anyone can be formed with conventional drilling tools—preferably brad point bits, which cut cleanly. The technique is like

8-30 The combination of fingers and square mortises results in a unique joint design. The fingers can project or they can be sanded flush.

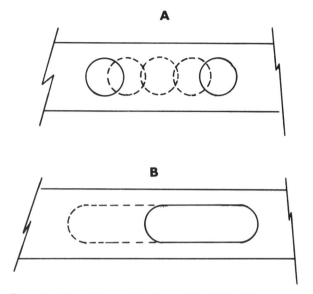

8-31   Round-end mortises can be formed by drilling (A) or routing (B).

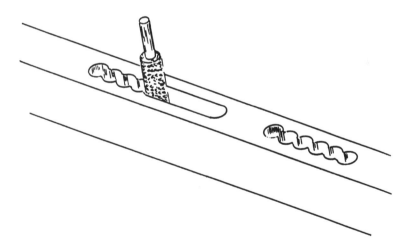

8-32   Remove the peaks that remain after overlapping holes are drilled by using a rotary rasp whose diameter matches the width of the mortise.

drilling holes on a common centerline. Select a bit with a diameter that matches the width of the mortise, then drill two holes that will determine the length of the mortise. The bulk of the waste between the initial holes is removed by making overlapping cuts (FIG. 8-31). The peaks that remain can be removed by hand with a chisel, or with a drill press mounted rotary rasp of suitable diameter (FIG. 8-32).

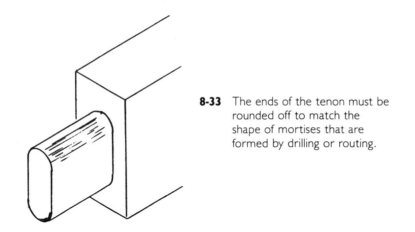

8-33 The ends of the tenon must be rounded off to match the shape of mortises that are formed by drilling or routing.

Mortising can also be accomplished with router bits, which will be covered in Chapter 9. Mortises formed by drilling or routing will have round ends, and the tenons must be shaped accordingly (FIG. 8-33).

TABLE 8-2 provides the solutions for some of the problems you might encounter when doing mortising work.

Table 8-2   Troubleshooting: Mortising

TROUBLE	CAN BE CAUSED BY	CHECK FOR OR DO
Overheating	RPM incorrect	Reset as close as possible to efficient speed
	Dull chisel bit	Sharpen cutting tools
	Waste accumulating in chisel or work	Retract frequently to clear chips—remove chips from slot in chisel—do not bury slot in work
	Misalignment	Check for correct clearance between chisel and bit
Tapered Mortise	Chisel leading off into first cuts	Keep cuts to 3/4 the width of the chisel
	Work movement	Keep work clamped securely
	Misalignment	Be sure table is in correct horizontal position
Serrated Edge on Mortise	Misalignment	Be sure side of chisel is square to fence
Excessive Feed Pressure Needed	Dull chisels/bits	Sharpen cutting tools
	Very hard wood	Drill relief holes first—try different speed
Work Lifts When Chisel retracted	Poor practice	Always use hold-downs or clamps to keep work secure
Wrong Depth	Poor practice	Check setting of drill press stop rod

Chapter **9**

The drill press as a router

Routing on a drill press is done with the same bits that are used with a portable router. The drill press is not meant to replace the router; it can't perform in areas that are exclusively in the realm of the portable machine. However, there are phases of the drill press application that offer some advantages. For example, it can situate and guide the workpiece while the cutting tool is in a stable position, adding to the convenience and accuracy of many operations. It's even possible that the big machine will permit using heavier and a greater assortment of bits. The router might be limited to gripping bits with a 1/4-inch shank (a common capacity), while the average drill press can hold cutters with up to 1/2-inch shanks. Thus, particular cutters that are not available in smaller sizes can be used.

SPEED

Conventional routing is done at speeds of 10,000 RPM and higher—speeds that can't be matched by drill presses where the common, maximum RPM can range from 3,000 to 5,000. Compensation is available in the form of a slower feed rate, which is how fast the work is moved past the cutter. If optimum results are produced by moving work at a particular speed past a cutter turning at 10,000 RPM, it seems reasonable that you can achieve similar results with a cutter turning at 5,000 RPM by reducing the feed rate 50 percent.

Don't change the pulley and belt assortment or add an intermediate pulley to get speeds higher than the tool's maximum without consulting the supplier or manufacturer. The design of the machine might not tolerate the modification.

A SPECIAL CHUCK

Because router bits develop considerable side thrust, they must be gripped in a special chuck instead of the three-jaw unit (FIG. 9-1). The chuck attaches to the spindle in a way that is peculiar to the machine it was designed for; it is not interchangeable. It is necessary to purchase a unit that's right for the tool you own.

If the available chuck will grip only 1/2-inch shanks, you can adapt it for 1/4-inch shanks by using a long, steel bushing with a 1/2-inch outside diameter and a 1/4-inch hole. The bushing is drilled and tapped for a couple of small set screws that will secure the cutter. The bushing, in turn, is gripped by the set screws in the chuck. If you can't find a suitable bushing or you don't care to make the accessory, have a local machine shop produce it.

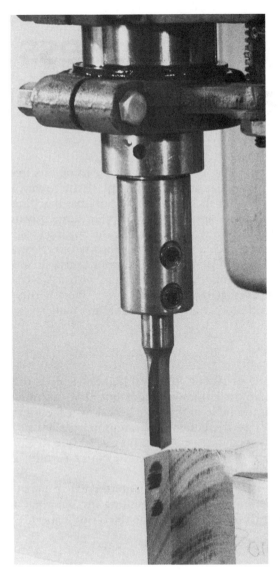

9-1   A special chuck is used for cutters that, like router bits, develop side thrust. Set screws instead of geared jaws grip the bit securely.

THE CUTTERS

Router bits come in a wide variety of shapes and sizes. Some are designed for practical applications like forming dadoes and grooves, rabbets and tongues, dovetails, and so on. Others, with particular cutting edge configurations, are mainly intended for fancy work like decorative edging on tabletops.

Regardless of their purpose, router bits all fall into one of the three categories shown in FIG. 9-2. Pilots bear against the work's edge and so control the width of the cut. Work that is done with bits that have integral pilots must be done carefully. Because the pilot rotates with the bit, it can cause enough friction to burn the work or the bit or both. The ball bearing pilot causes no such problems because it turns independently, matching the speed at which the work is moved. Piloted bits can be used in a drill press either by allowing the pilot to guide the work or by using a fence.

Straight bits do not have pilots. When they are used in a drill press, the work *must* be guided by a fence.

TWO ROUTER JIGS

In order to be prepared for all phases of drill press routing, it's necessary to make a special jig that can be secured to the machine's table. The jig consists of an auxiliary table that is grooved to receive a miter gauge, and an adjustable fence that serves as a guide when doing straight-line routing. Because this type of accessory is not available commercially, it should be a first project. Do it before you move into the routing phase of drill press applications.

The fence for the jig detailed in FIG. 9-3 is a two-piece design that can provide a smooth bearing surface for the work when straight bits are used for jobs like grooving. Or it can be separated to provide a center gap when only part of the cutter must contact the work. The position of the jig's table determines the distance from the fence to the cutter. A single C-clamp at each side of the jig is used to secure the positions of both table and fences.

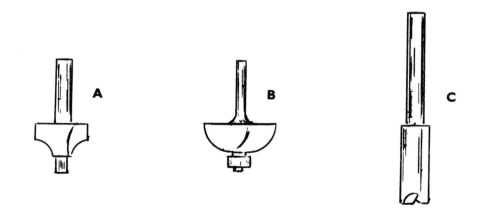

9-2 Integral pilot bits turn at the same RPM as cutting blades (A), ball bearing pilot bits turn independently of blades (B), and straight bits do not have pilots (C).

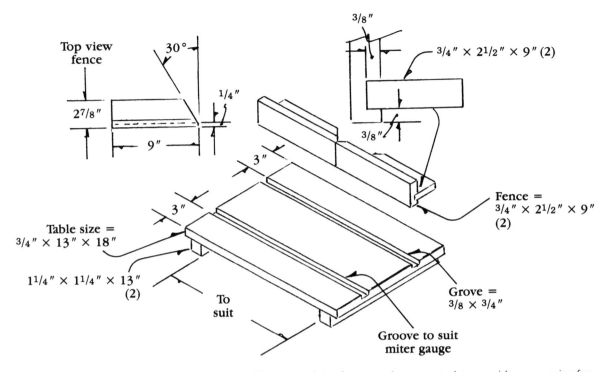

9-3 Construction details for router jig No. 1. The parts of the fence can be separated to provide an opening for cutters.

The project shown in FIG. 9-4 and detailed in FIG. 9-5, is another version of a router jig. In this case, the fence is a one-piece design that has a relief area at its center point to accommodate straight or profile bits. A convenience factor is that the table is locked in a permanent position. The adjustable fence is secured to the table with a C-clamp at each end.

Both designs call for a miter gauge groove for safe guidance of narrow work that must be end-shaped. The groove can be sized for a miter gauge that you have on hand or buy for the purpose. The important consideration is that narrow stock must never be fed past a cutter without using a backup.

SAFETY

The critical safety consideration is accepting that cutters can hurt you. Never work so that your hands come close to one. Don't place your hands on line with a cutter when you are feeding stock, and never try to rout stock that is too small for a smart hand position.

Make a guard that provides a safety shield (FIG. 9-6). The shield, which is adjustable vertically so it can be used with various stock thicknesses, is adjusted so there is minimum clearance between it and the surface of the work. The shield is not shown mounted in the illustrations but it is left off only so cuts can be seen more clearly. Information for making the guard was offered in Chapter 4.

9-4 Router jig No. 2. The one-piece fence on this unit has a center cutout for bits. The miter gauge has a hold-down that automatically presses down on the work when the handle is squeezed.

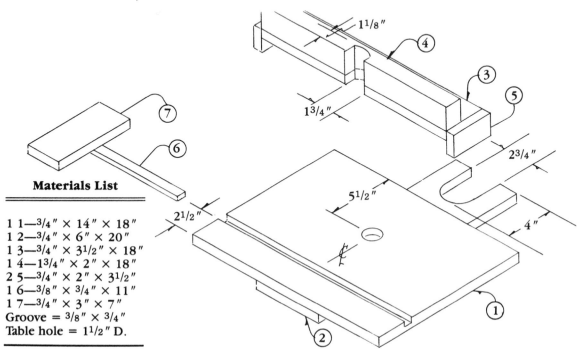

Materials List

1 1—$^3/_4$" × 14" × 18"
1 2—$^3/_4$" × 6" × 20"
1 3—$^3/_4$" × 3$^1/_2$" × 18"
1 4—1$^3/_4$" × 2" × 18"
2 5—$^3/_4$" × 2" × 3$^1/_2$"
1 6—$^3/_8$" × $^3/_4$" × 11"
1 7—$^3/_4$" × 3" × 7"
Groove = $^3/_8$" × $^3/_4$"
Table hole = 1$^1/_2$" D.

9-5 Construction details for router jig No. 2. The jig is held in place with a large clamp that squeezes the U-shaped opening against the drill press column. Note the design for a simple miter gauge.

9-6 Construction details for the safety shield were shown in Chapter 4. It should be used for all routing operations even though it isn't evident in the photos.

PROCEDURE

Use the highest RPM that the machine can provide, and adopt a feed speed that allows the cutter to keep working. Keep quill extension to a minimum and be sure its position is securely locked. Always move the work *against* the cutter's direction of rotation. On straight cuts (router bits in a drill press rotate clockwise), this means passing the work from left to right (FIG. 9-7).

Straight cuts are made by using the fence for guidance and support. The left-to-right feed direction utilizes the cutter's rotation to help keep the work in position. The action of the cutter will tend to move the work away from the fence if the feed direction is reversed (FIG. 9-8).

Keep the work tightly down on the table and maintain firm contact with the fence throughout the pass (FIG. 9-9). Don't harass the cutter and the machine by trying to force deep cuts. When necessary, use a repeat pass technique to get the job done. For example, if you need a groove or a dado that is 1/2 inch deep, form one to half depth and then cut again after resetting the cutter.

Rabbeting bits have a cutting diameter that is greater than the shank and are used, naturally, to produce the L-shaped rabbet (FIG. 9-10). The same cutter can be used to shape tongues (for tongue-and-groove joints) by making a second pass

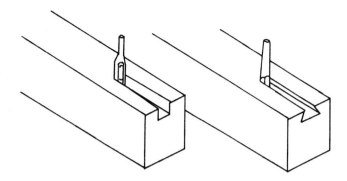

9-7 Router bits turn in a clockwise direction, so work feed should be from left to right.

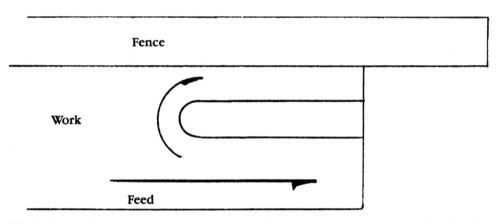
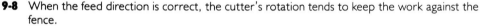

9-8 When the feed direction is correct, the cutter's rotation tends to keep the work against the fence.

after the stock has been inverted. Whenever possible, here and in other situations, try to set up so the cutter is *under* the workpiece. Then, should you accidentally lift the work during the pass, the cutter won't dig in as it would if it was above the work.

Shaped edges are produced by making straight-line cuts with the fence and cutter height organized for what is needed. If the jig has a two-piece fence, set them so there will be a minimum clearance for the bit. Never, as already cautioned, try to shape the end of narrow pieces without using a suitable backup (FIG. 9-11). The work will not have enough support surface for safety, especially when it moves past the opening around the cutter, if you don't provide the correct means for moving it.

CURVED WORK

Rout components that can't be guided by a fence using piloted bits, preferably those of ball bearing design (FIG. 9-12). A major factor for safety and good results is to be sure there is sufficient bearing surface between the work and the pilot. The

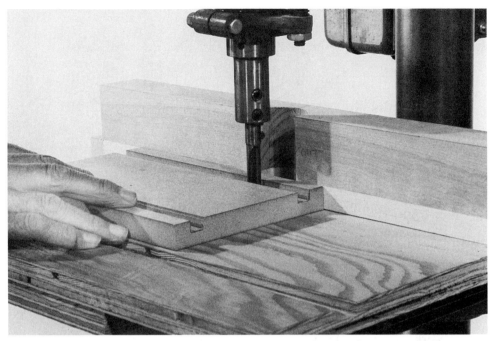

9-9 Work must be firmly on the table and snug against the fence. Don't feed the work any faster than the bit can handle. Make repeat passes, when necessary, to achieve full cut depth.

9-10 Router bits can be used to form rabbets. Invert the work and make a second pass and you have a tongue. The arrow indicates the feed direction.

9-11 End-cutting must always be done with a backup, preferably a miter gauge. It isn't safe to do this kind of work without correct work support.

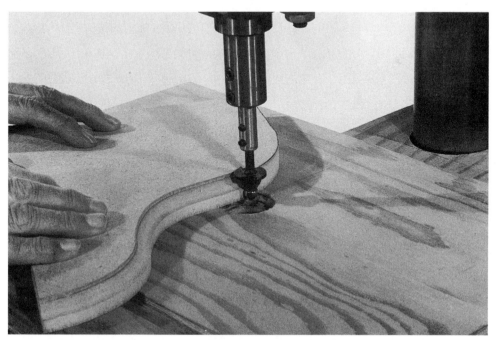

9-12 Curved edges can be shaped by using a piloted router bit. The work bears against the pilot throughout the pass. Note the homemade adapter that allows using router bits with a 1/4-inch shank in a 1/2-inch chuck.

work must be moved *against* the cutter's direction of rotation. If you try to operate otherwise, the cutter will surely pull the work from your hands.

Keep a firm grip and move the work steadily. Be sure, before you do the routing, that the work's edge is perfectly smooth. Any irregularity will simply be followed by the pilot. Inside cutouts can be handled the same way. Lower the cutter and set it for correct height after the work is placed on the table (FIG. 9-13).

PIVOT ROUTING

Circular shapes can be routed by using the pivot guide method on the routing jig or by clamping a platform directly to the drill press table (FIG. 9-14). The distance from the pivot point, which can be a nail, to the cutter is the radius of the circle. Hold the work firmly, or clamp it when you lower the bit to start the cut, then rotate the work in a counterclockwise direction. This is a good technique to use when you need a hole or a disc that is beyond the capacity of a fly cutter. Just make repeat passes until the cutter has penetrated the workpiece.

Figure 9-15 shows how to use a jig to form grooves in the edge of rings or segments of a circle. The contours of the jig and the work must match.

9-13 Inside cutouts can also be shaped using the piloted router bit. Put the work in position before you lower the bit. Always be sure the quill position is secured before you turn on the machine.

9-14 Use a pivot jig to form a circular groove. Work feed is clockwise. Use this method to produce oversize holes or discs.

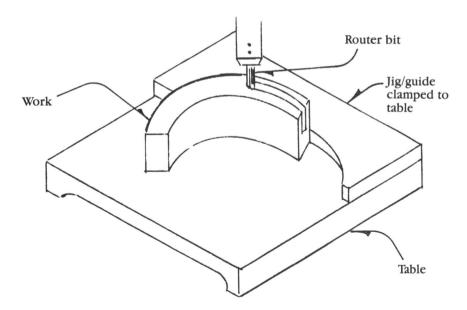

9-15 Routing a groove in the edge of a curved piece. The idea will work only if the shapes of jig and work are compatible.

9-16 Routing a mortise. Use the machine's depth stop rod to control the depth of the cut.

ROUTING MORTISES

Straight-shank bits can be used to form smooth and accurate round-end mortises (FIG. 9-16). Because straight bits are available in various diameters, it's probable there will be one that will produce the mortise width that is needed. Mortises are usually deep, so accept that you will have to make repeat passes to get the job done. Don't forget that the pass must be made from left to right. This means that each cut must start at the right end of the mortise.

Open-end mortises are formed in the same manner, except that the work is moved directly into the cutter (FIG. 9-17). In both cases, you can clamp a stop block to the table or the fence to gauge the length of the cut.

Mortises formed this way can be squared off by hand with a chisel. However, it's easier if you round off the mating tenon to conform with the mortise, and it won't lessen the joint strength.

FORMING DOVETAILS

The multiple cuts that are required for the conventional dovetail joint are best formed with a portable router and a commercial jig made for the purpose. How-

9-17 Routing an open-end mortise. The work is moved directly into the cutter. Repeat passes are usually needed when forming mortises. A clamped stop block will gauge the length of the cut.

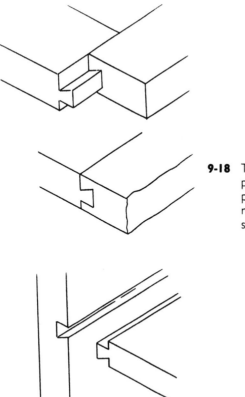

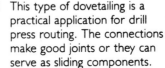

9-18 This type of dovetailing is a practical application for drill press routing. The connections make good joints or they can serve as sliding components.

ever, slide-type dovetails are practical applications for drill press routing (FIG. 9-18).

Dovetail slots in surfaces are formed by duplicating the setup shown in FIG. 9-19. This is straight-line work, but because the shape of the bit must be "cloned," you can't use repeat passes to achieve full depth. However, the following technique is a solution if you find it difficult to do the job in a single pass. First use a straight bit, the diameter of which matches the narrow end of the dovetail bit. Run a groove down the centerline of the cut. Switch to the dovetail bit and make a second pass to complete the job. At that point, there will be little material for the dovetail bit to remove. Dovetails that are required in the edge of components are formed by using the setup shown in FIG. 9-20.

The matching dovetail "pin" requires two passes. Organize the cutter and the jig's fence so the first pass will shape one side of the pin. The form is completed by making a second pass after the work has been turned end-for-end (FIG. 9-21). Obviously, the care used to create the setup will determine the cut's accuracy. Take the precaution of testing on a piece of scrap material before working on good stock.

V-BLOCK WORK

The joint used to connect legs to a pedestal is often a dovetail. Forming the dovetail in the pedestal can be fairly straightforward if you make the V-block jig shown

9-19   Forming dovetails in surfaces is a straight-line routing chore.

9-20   Dovetailing an edge. Jobs like this can be done more easily if you clamp a strip of wood to the table as a second fence. The two fences will form a neat channel for the work, assuring its vertical position throughout the pass.

9-21 The dovetail pin is shaped by making a similar cut on both sides of the stock. One of the cuts should be made very slowly because it will be against the grain. Test the setup for accuracy before you cut into good material.

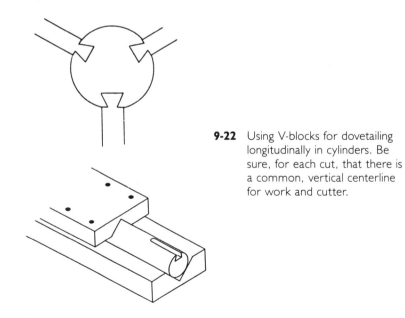

9-22 Using V-blocks for dovetailing longitudinally in cylinders. Be sure, for each cut, that there is a common, vertical centerline for work and cutter.

in FIG. 9-22. The jig, with the workpiece firmly captured, is moved along the fence just as if a simple straight cut was involved. Clamp a stop block to the fence to control the length of the cut.

Check TABLE 9-1 if you encounter problems when doing drill press routing.

Table 9-1   Troubleshooting: Routing

TROUBLE	CAN BE CAUSED BY	CHECK FOR OR DO
Tapered Cut	Misalignment	Check for table squareness to spindle
	Work problem	Bearing edge of work must be square to sides
Work Pulls from Fence	Poor practice	Feed work *against* cutter's rotation—keep work snug against fence
Work Hits Edge of Outfeed Fence	Misalignment	Bearing surfaces of fence must be on same plane
Wrong Depth	Setup	Check depth stop setting
Hard to Feed	Feed direction	Work *against* cutter rotation
	RPM	Use fastest speed
	Deep cut	Use repeat pass technique
Feathering/Splintering on Cross-Grain Cuts	Characteristic	Reduce feed speed toward end of cut—do cross-grain cuts first, end with with-the-grain cuts
Chatter	Cutting too deep	Use repeat passes
	Work handling	Keep firm grip—keep work snug on table and against fence
Burn Marks	Dull cutter	Sharpen or discard
	Cutting too deep	Use repeat pass technique
	Feeding too fast	Maintain consistent feed speed and pressure
	Feeding too slow	Maintain consistent feed speed and pressure

Chapter 10

The drill press as a shaper

Shaping is a woodworking application used to produce decorative edges on furniture components and to form a variety of joints. While the drill press is not as flexible as an individual shaping machine, it will serve for many chores that would otherwise require an additional, expensive tool. Some major differences between the tools have to do with spindle position and speeds. A shaper has its spindle under the table and works with speeds that are, on average, twice the maximum of RPM of a drill press.

Shaper-cutters used in a drill press attach by means of an adapter to the spindle and are positioned above the table. The shaper needs no assembly. The drill press needs a special table accessory that might be available for the tool you own. You can also make one (FIG. 10-1).

You can compensate considerably for reduced speed by how fast you feed the work. This was discussed in relation to drill press routing, Chapter 9.

Having the spindle above the table is not critical for most shaping operations, so this factor is negligible.

ADAPTER

An adapter that replaces the three-jaw chuck is used to secure the special cutters that are required for shaping cuts. Its design will depend on the machine it was designed for, but it will have a grooved, threaded shaft, a lock nut, and a keyed washer that is placed under the nut to prevent it from loosening. The shaft will be long enough to accommodate the cutter, plus ''collars'' that serve various purposes (FIGS. 10-2 and 10-3).

Collars are available in various thicknesses and diameters. Their main functions are to control width-of-cut when shaping is done without a fence (freehand

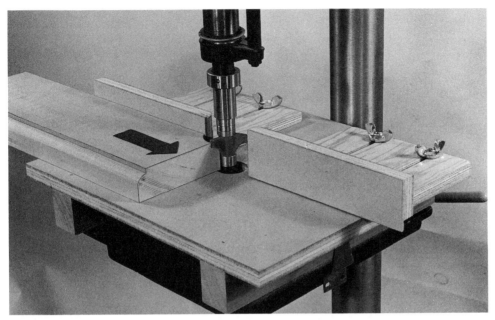

10-1   Drill press shaping requires a special table with individually adjustable fences. You can make one like this if a commercial one is not available.

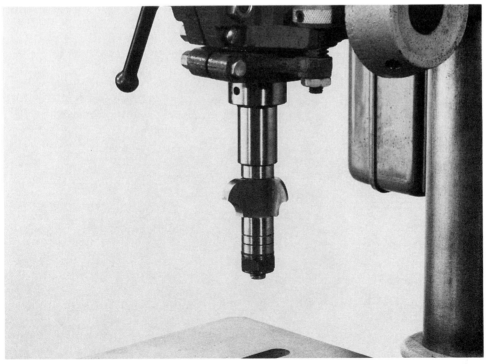

10-2   A special adapter that replaces the standard three-jaw chuck is used for drill press shaping.

I0-3 These are typical accessories that are required for shaping.

shaping), or to position a cutter at a particular point on the adapter. They are used over, under, or even between cutters, depending on the work being done (FIG. 10-4). Their use is optional when a fence is used for control during straight-line shaping, but they *must* be used for freehand shaping. A few collars will probably come with the adapter, but it's wise to add others of different diameters since they have an important role in shaping operations. Maintain them in sparkling condition. Workpieces that bear against them will be marred if the collars are not smooth, or are dirty or scratched.

THE CUTTERS

Three-lip shaper-cutters that are standard equipment for individual machines are practical for use with a drill press (FIGS. 10-5 and 10-6). They are rugged tools with a lot of backup for cutting edges, so they will stay sharp for long periods if they are not abused. Those that are available with carbide-tipped blades are maximum performers in terms of cut smoothness and sharpness longevity. They are more expensive than all-steel types, but are worth it—especially when used for production output.

Shaper-cutters are available in a variety of shapes that fall into the category of combination or full profile. Combination types are intended for partial cuts. That is, just a portion of the cutting profile may be used to produce a particular form (FIG. 10-7). The form is controlled by the cutter's position above the table and the width of the cut, which is determined by fence position or collar diameter. Usually combination units are used for decorative work.

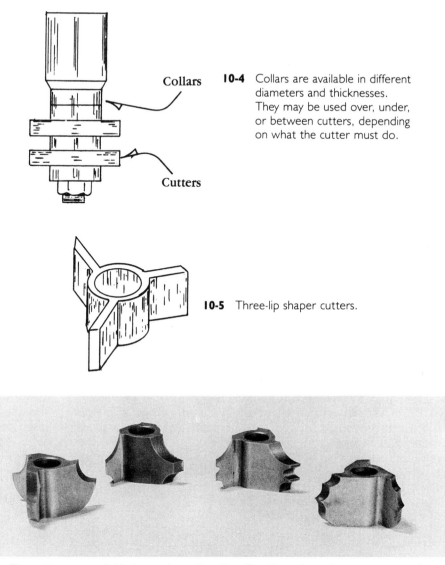

Collars

10-4 Collars are available in different diameters and thicknesses. They may be used over, under, or between cutters, depending on what the cutter must do.

Cutters

10-5 Three-lip shaper cutters.

10-6 The cutters are available in a variety of profiles. The three-lip versions are commonly used on individual shaping machines and are the tools to use for drill press work.

Full profile cutters are meant to produce shapes, such as joints, that have practical applications (FIG. 10-8). Cutters are also available in matched pairs to produce mating forms. Those shown in FIG. 10-9 are for tongue-and-groove joints. Keep in mind that you can use part of *any* cutter's profile if it results in the shape you are seeking.

Cutters must be treated with respect when they are stored as well as when they are in use. Keep them from nicking each other and from being blunted by other tools. The project in FIG. 10-10 offers a storage solution: a simple case with a hinged lid and a bottom that is studded with dowels to keep the cutters separated.

Full cut   Cutter profile

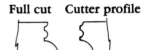

**Forms produced by using
only part of the cutter's profile**

10-7 Combination cutters are usually used for partial cuts. Thus the one cutter can provide various contours. Controls are: the height of the cutter in relation to the work's edge, and the depth of the cut.

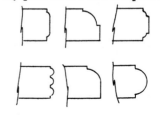

Combination types

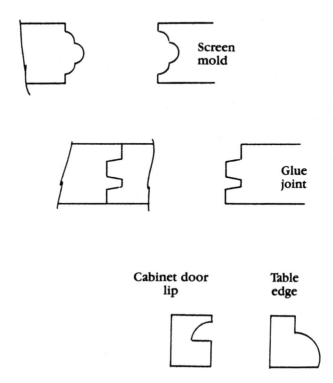

Screen mold

Glue joint

Cabinet door lip   **Table edge**

10-8 Some cutters, like these, are intended for full profile shapes.

Tongue and groove
(set)

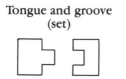

10-9 Some cutters are available in
pairs so each produces a form
that suits the other one.

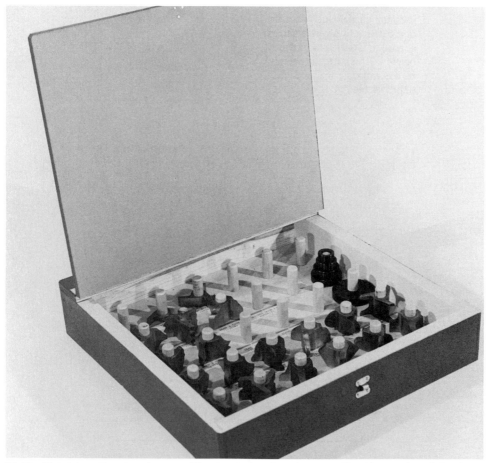

10-10 Maintain cutters in prime condition by storing them in a case like this one. Install the dowels so the cutters will have adequate separation.

SHAPING JIG

Shaping requires a special table and fences that are specifically designed for that phase of drill press work. It is not the most common accessory but some manufacturers, like Shopsmith, offer one that is just right for the machine they sell (FIG. 10-11). It's possible that a table for one machine could be adapted for use on a rival

10-11 Shopsmith offers a shaper jig accessory that is used when the machine is in its vertical drill press mode.

product but this is a risky option. The solution is to make your own, like the one shown in FIGS. 10-12 and 10-13. The project is practical and emulates the features found on professional units: twin fences that are individually adjustable to-and-fro to control depth of cut, and that can be moved laterally to minimize the opening

10-12   This shaper jig that you can make has the features of a professional unit. Fences are individually adjustable to-and-fro and laterally to minimize the opening around the cutting tool.

around the cutting tool. The slots in the fences can be formed with straight router bits.

The need for individually adjustable fences is explained in FIG. 10-14. The fences are set for a single bearing plane when only part of the work edge is removed. When the entire edge of the stock is cut off, the outfeed fence is brought forward so the work will be adequately supported after it has passed the cutter.

PROCEDURE

Use the machine's highest speed and a work feed pressure that is only enough to keep the cutter working. Forcing will cause overheating and rough edges, if not excessive slowing of the motor or a complete stall. On the other hand, being too cautious won't accomplish much. Always move work *against* the cutter's direction of rotation, which, as in routing, means moving from left to right. Whenever possible, operate so that the cutter is under the work. This will afford some protection from the cutter and will prevent marring, should you inadvertently allow the work to raise during the pass.

Keep your hands away from the cutting area and don't place either one of them on the cut path. *Never* try to shape small pieces. If you must shape a small

Table

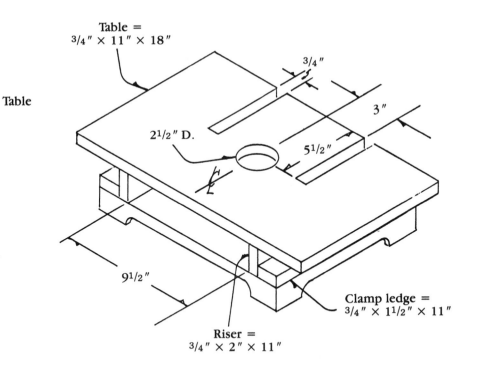

Table =
3/4" × 11" × 18"

3/4"

3"

2¹/₂" D.

5¹/₂"

9¹/₂"

Riser =
3/4" × 2" × 11"

Clamp ledge =
3/4" × 1¹/₂" × 11"

Fences

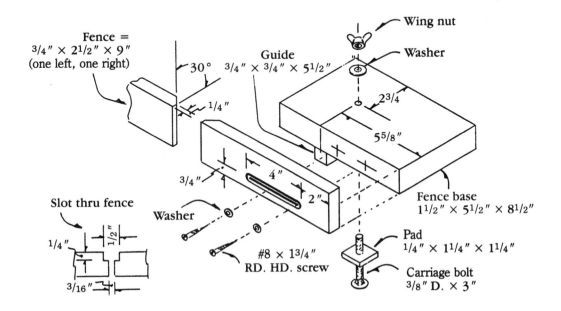

Fence =
3/4" × 2¹/₂" × 9"
(one left, one right)

30°

1/4"

Guide
3/4" × 3/4" × 5¹/₂"

Wing nut

Washer

2³/₄

5⁵/₈"

3/4"

4"

2"

Slot thru fence

1/4"

1/2"

3/16"

Washer

#8 × 1³/₄"
RD. HD. screw

Fence base
1¹/₂" × 5¹/₂" × 8¹/₂"

Pad
1/4" × 1¹/₄" × 1¹/₄"

Carriage bolt
3/8" D. × 3"

10-13 Plans for the drill press shaper accessory. Check the dimensions against the machine you own before cutting the components.

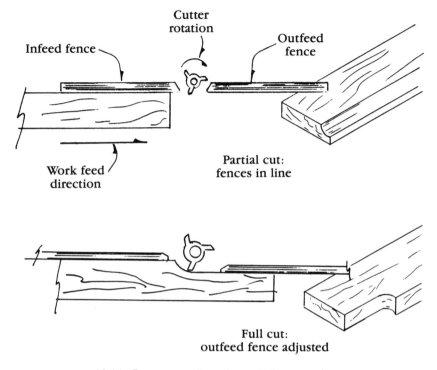

Cutter
rotation

Infeed fence

Outfeed
fence

Work feed
direction

Partial cut:
fences in line

Full cut:
outfeed fence adjusted

10-14 Fences are adjusted to suit the operation.

piece, do the job on stock that is safe to handle and then use another machine to cut off the section you need. Use the homemade safety shield described in Chapter 9, even though it is not present in the illustrations in this chapter.

STRAIGHT LINE SHAPING

Straight edges are always shaped using the fences for guidance and support. The position of the fences in relation to the cutter determines how wide the cut will be. The cutter's height, controlled by quill extension, sets the cut depth. Adjust the fences so the gap between inboard edges is only enough to supply clearance for the cutter. Do not turn on the machine until you are sure the fences are locked and that the jig is firmly clamped to the drill press table.

Keep the work down on the table and snug against the infeed fence as you start the cut. Then advance it smoothly past the cutter (FIG. 10-15). If the work is stopped by the outfeed fence or if it lacks support after passing the cutter, you will know that you did not adjust the fences carefully.

To help move the work smoothly, clamp a strip of wood to the infeed fence so it will serve as a hold-down. Set it so it barely touches the surface of the work; if it forces the work down on the table it will be difficult to move (FIG. 10-16).

Blank cutters are used for slight jointing cuts or to form L-shaped rabbet. The same setup, with cutter height adjusted, will form a tongue when a second pass is made after the work has been inverted. For example: If a $1/4$-inch-thick tongue is

10-15 Work is held for firm contact down on the table and against the fence. Move the work steadily and at a pace that allows the cutter to work without being forced. This is a full cut so the outfeed fence has been brought forward to support the work after it has passed the cutter.

10-16 A strip of wood clamped to the infeed fence serves as a hold-down, so you need be less concerned about inadvertently allowing the work to lift during the pass.

needed on ³/₄-inch stock, the cutter is set for a cut that is ¹/₄ inch deep. The position of the fences will determine how wide the tongue will be.

GRAIN DIRECTION

Smoothest cuts are achieved when the work is moved with the wood grain. Against-the-grain passes will always result in some tearing at the end of the cut. When shaping is required on adjacent edges or all four edges of a component, do the cuts that are against the grain first. The final cuts made with the grain will eliminate the imperfections caused by the first passes (FIG. 10-17).

SLIM MOLDINGS

Drill press shaping can be used to produce slim moldings or other narrow components safely, but only if the procedure shown in FIG. 10-18 is followed. The idea is to work with stock that is large enough so your hands can be well away from the cut area. After the first cut, saw the shaped edge off on another machine and continue the process until you have the number of parts you need. The sawing should be done with a smooth-cutting blade so the new edge will bear nicely against the shaper-jig fences.

FREEHAND SHAPING

Freehand shaping is used when components must be formed on inside or outside curved edges (FIG. 10-19). Because the work can't be supported and guided by a fence, a special system is used to do the job accurately and safely. The system includes using collars together with the cutter on the shaper adapter (FIGS. 10-20 and 10-21), and fulcrum pins against which the work can be braced before and during the cut.

A basic setup that provides the fulcrums is shown in FIG. 10-22. The pins, which can be made by removing the heads from steel bolts or by threading a portion of a steel rod, are positioned on the table by means of threaded sleeves. The

10-17 Sequence of passes when shaping four edges. Make cross-grain or against-the-grain cuts first.

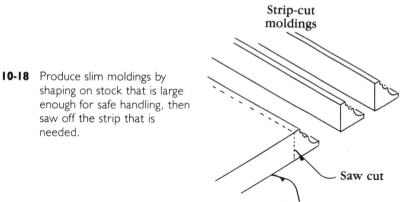

10-18 Produce slim moldings by shaping on stock that is large enough for safe handling, then saw off the strip that is needed.

Strip-cut moldings

Saw cut

Base stock

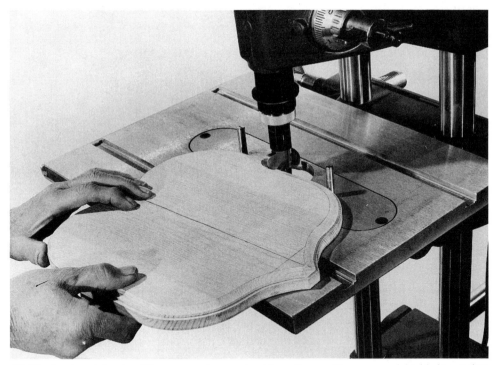

10-19 Freehand shaping. The accessories required for a Shopsmith are a special table insert that is threaded for fulcrum pins.

same idea can be adopted for the shaper-jig (FIG. 10-23). If the threaded portion of the pins is long enough, a nut can be used under the table to hold the pins even more tightly.

The guard shown in FIG. 10-24 should be made and used whenever the jig is set up for freehand shaping. The guard's position is adjusted so it fits nicely over

10-20 Collars alone support the work during freehand shaping. In order for the work to go smoothly, the collars must be maintained in pristine condition.

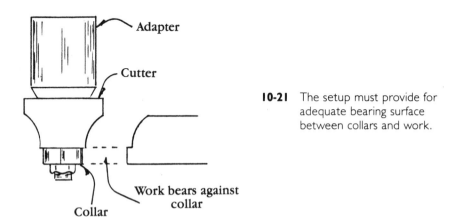

10-21 The setup must provide for adequate bearing surface between collars and work.

the fulcrum pins while providing for the shaper adapter and cutters. It will keep your hands away from the cut area without hiding what is going on. Make the guard following the plan in FIG. 10-25.

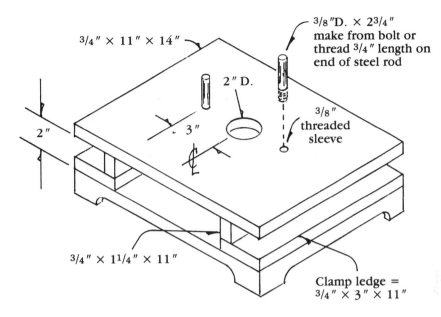

3/4″ × 11″ × 14″

2″ D.

2″

3″

3/8″D. × 2³/4″
make from bolt or
thread ³/4″ length on
end of steel rod

3/8″
threaded
sleeve

3/4″ × 1¹/4″ × 11″

Clamp ledge =
³/4″ × 3″ × 11″

10-22 Plans for a basic jig that can be designed for the use of fulcrum pins.

10-23 Equipping the shaper jig with fulcrum pins makes it a complete shaping accessory.

10-24 This freehand shaping guard is designed for use with the shaper jig.

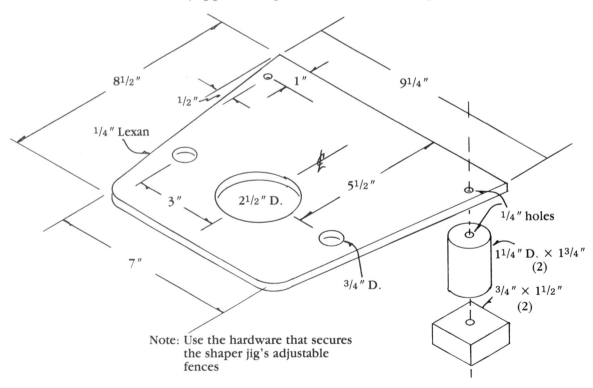

8¹/₂″

1/2″

1″

9¹/₄″

¹/₄″ Lexan

3″

2¹/₂″ D.

5¹/₂″

7″

3/4″ D.

¹/₄″ holes

1¹/₄″ D. × 1³/₄″
(2)

3/4″ × 1¹/₂″
(2)

Note: Use the hardware that secures
the shaper jig's adjustable
fences

10-25 Plans for the shaper jig guard.

10-26 Freehand cuts are started by bracing the work against the infeed pin and then slowly advancing it until it contacts the cutter and rests firmly against the collars.

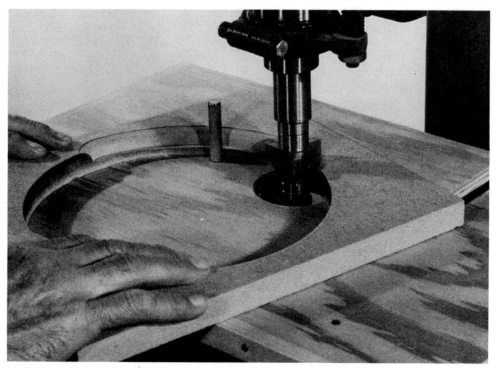

10-27 Positioning the work before lowering the cutter allows shaping on internal cutouts. Work is moved counterclockwise. (The guard is removed for illustration only.)

Start freehand shaping by bracing the work against the infeed (left side) fulcrum pin and then advancing it very slowly until it contacts the cutter and seats firmly against the collars (FIG. 10-26). There can be occasions when the size or the shape of the work will make it necessary to swing away from the pins, especially the outfeed pin. In such cases, bear in mind that the collars alone are providing work support. Be cautious at the end of all cuts to be sure the work will break the cutter contact smoothly.

INTERNAL CUTOUTS

Use the freehand technique when you wish to shape edges of forms that are contained within a workpiece. This is possible because the work can be placed in position before the height of the cutter is established. Shaping then proceeds in fairly routine fashion, with firm contact between work and collars and with work support supplied by one or both of the fulcrum pins (FIG. 10-27). Work feed, as always, is *against* the cutter's direction of rotation. If you try to do otherwise, the work will surely be pulled from your hands.

Always maintain a firm grip on the work and be especially careful at those times when the contours of the work may preclude the use of fulcrum pins. Use the guard! If the mounting arrangement for the freehand shaping guard interferes because of work size or shape, use the vertical safety shield instead.

Don't force cuts. Make repeat passes when necessary. This operation works for contoured edges as well as for straight cuts.

TABLE 10-1 lists problems that might arise when shaping, and possible solutions.

Table 10-1 Troubleshooting: Shaping

TROUBLE	*CAN BE CAUSED BY*	*CHECK FOR OR DO*
Cuts Rougher Than Should Be	Rushing the cut	Pass the work more slowly
	Feeding against grain	Make pass with the grain whenever possible
	Dull cutting tool	Keep cutters sharp
	Cut is too deep	Use repeat pass technique when necessary
	Wrong RPM	Use fastest speed
	Poor practice	Keep work snug against fence or collars throughout pass
Feathering or Splintering on Cross-grain Cut	Characteristic	Do cross-grain cuts first, finish with the grain—use scrap block as backup at end of cut
Collar Marks on Work	Collars marred	Replace
	Collars gummed up	Clean and store carefully
Work Tends to Pull From Hand	Wrong feed direction	Always move work against cutter rotation
Kickback	Poor practice—lack of support	Start cut slowly—keep work against fence—always use fulcrum pins when freehand shaping
Burn Marks	Dull cutter	Sharpen
	Cutting too deep	Use repeat pass technique
No Support from Outfeed Fence	Poor practice	Keep fences aligned when making partial cuts—adjust outfeed fence when making full cut
Depth of Cut not Uniform	Misalignment	Check fence positions
	Poor practice	Keep work snug against fence or collars and down on table throughout pass
Work Gouged	Poor practice	Keep work firmly down on table
	Cutter position	Try to work so cutter is under the work
Bumps on Shaped Edge	Work pressure variation	Keep firm grip and steady pressure-work must be snug against fence or collars and down on table throughout pass

Chapter **11**

Drum sanders

The drill press is an excellent power tool for drum sanding chores. A major advantage is that the tool's table can provide support for workpieces. Also, because the angle between the table and a spindle-mounted drum is 90 degrees, a sanded edge will be square to adjacent surfaces.

Some workers doing drum sanding will swing the table aside and lower the drum so its bottom edge is below the table's surface. Thus, work resting on the table can be brought to bear against the abrasive surface. This is a simple and acceptable method. However, to increase support so it's easier to keep the work level, and to be able to place work at any point around the drum, try to provide an auxiliary table (FIG. 11-1).

THE DRUMS

Sanding drums are available in various diameters and lengths (refer to FIG. 3-13). Common sizes start at about 1/2 inch and range up to 2 1/2 inches. If you have an assortment you will be prepared to do edge sanding on, for example, straight edges, circles and curves, and inside rounded corners. The largest shaft on the tools is 1/2 inch, so gripping them in the standard three-jaw chuck is no problem. Because the abraders do not develop side thrust when they are used efficiently—that is, without excessive force—special adapters or chucks are not needed.

Abrasive sleeves on most drum sanders are mounted or removed by means of a nut that is at the bottom or top end of the drum. Turning the nut one way or the other causes the rubber cylinder to expand or contract, either gripping firmly or relaxing its pressure against the sleeve.

The rubber cylinders are sturdy, but extended use can cause some distortion. Check the cylinders for surface uniformity (FIG. 11-2). If the problem is extremely slight, correct it by mounting the naked drum in the drill press and smoothing it

11-1 Drum sanding is a super application for a drill press. Sanded edges are guaranteed square to adjacent surfaces because the work is on a horizontal plane throughout the pass.

11-2 The cylinders should have flat surfaces.

with very fine sandpaper wrapped around a block of wood. This must not be overdone because a cylinder with a reduced diameter will not grip the sleeves.

ABRASIVE SLEEVES

Abrasive sleeves are available in various grits and minerals. Common grit designations are fine (#80−100), medium (#50−80), and coarse (#40−60). Most popular abrasive minerals are garnet and aluminum oxide. Both are acceptable for sanding wood. Aluminum oxide is generally considered preferable for sanding plastics and metal as well as wood.

The grit used will depend on the condition of the wood. For most procedures it is best to start with a coarse grit and then go to medium and fine. If the work does not require thorough sanding, start with a medium or even a fine grit.

JIGS TO MAKE

A basic drum sander jig is shown in FIGS. 11-3 and 11-4. Two specifics apply to jigs of this kind: The hole through the table should be about ¹/₈ inch greater than the diameter of the drum, and the worktable should be above the tool's table a distance that equals the length of the drum. This way the drum is free to move vertically its entire length so that all of its abrasive surface can be used.

The jig can be modified to provide maximum support when small sanders are used (FIG. 11-5). A support block that is in line with the drum's access hole is secured to the underside of the table. The extra part will then provide support for inserts, all of which have the same outside diameter, but with a center hole that suits the drum that will be used. Thus, there will be adequate support close to the drum even when small pieces are sanded.

11-3 The basic drum sander jig provides a raised platform so the drum can be moved vertically. This makes it possible to utilize the entire abrasive surface.

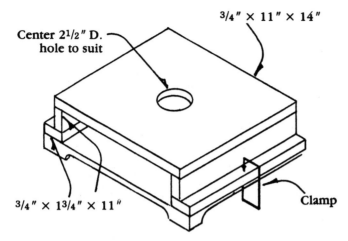

Center 2¹/₂″ D.
hole to suit

3/4″ × 11″ × 14″

3/4″ × 13/4″ × 11″

Clamp

11-4   Plans for a basic drum sander jig.

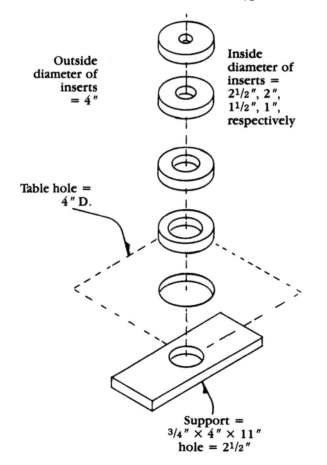

Outside
diameter of
inserts
= 4″

Inside
diameter of
inserts =
2¹/₂″, 2″,
1¹/₂″, 1″,
respectively

Table hole =
4″ D.

Support =
3/4″ × 4″ × 11″
hole = 2¹/₂″

11-5   Method of providing inserts for small-diameter drums, in order to give maximum work support around them when in use.

Figure 11-6 offers another version of the basic jig. In this case, the back area of the table is broadened so a long fence, which comes into service on some sanding applications, can be clamped in place. The jig in FIG. 11-7 has a pivoting action for adjusting its distance from the drum. It is secured with a single clamp.

The pivot action fence design was adopted for the jig that is offered in FIGS. 11-8 and 11-9.

A fence can also be used to provide guidance and support for a component that needs holes on a common centerline (FIG. 11-10).

SANDING CHORES

The grit of the sandpaper has some bearing on the speed at which the drum should be turning. Coarse papers will cut efficiently at speeds of between 1250 and 1500 RPM. You can use higher speeds with fine abrasives, but it is never necessary to exceed 2000 RPM. It won't take long before you can judge the right combination of abrasive grit and drum speed for the job at hand.

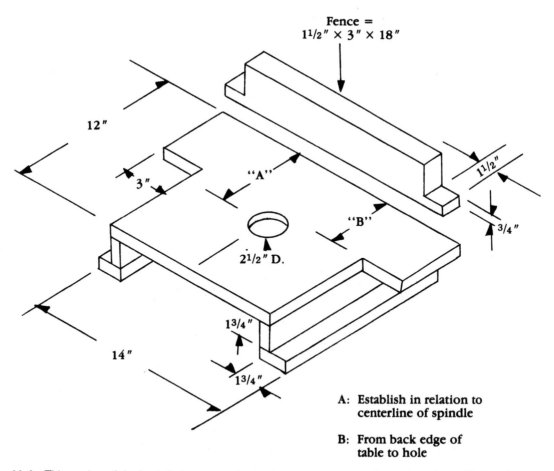

Fence =
1¹/₂″ × 3″ × 18″

12″

3″

"A"

1¹/₂″

3/₄″

"B"

2¹/₂″ D.

14″

1³/₄″

1³/₄″

A: Establish in relation to centerline of spindle

B: From back edge of table to hole

11-6 This version of the basic jig has a broader back area to accommodate a long fence. Reduced areas at each end of the fence allow the jig to be secured with a pair of C-clamps.

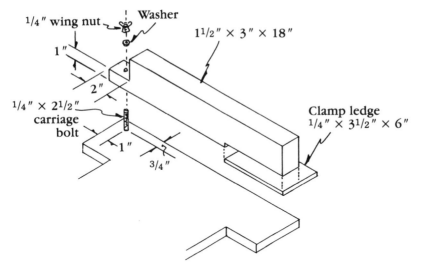

11-7   Alternate fence design with pivoting action for controlling the distance from the drum, and a ledge so that a single C-clamp will secure it.

11-8   Drum sander jig No. 2. The platform and the beam that supports it are U-shaped at the rear to conform with the drill press column. The concept incorporates a pivoting fence.

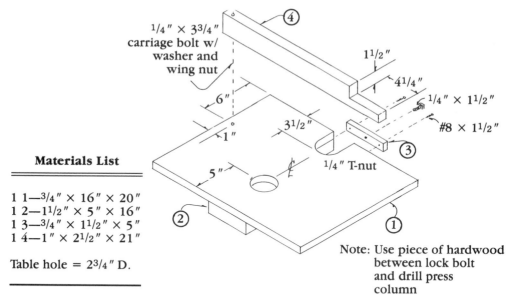

$1/4'' \times 3^3/4''$
carriage bolt w/
washer and
wing nut

$1^1/2''$

④

$4^1/4''$

$6''$

$1/4'' \times 1^1/2''$

#8 × $1^1/2''$

$3^1/2''$

③

$1''$

Materials List

$1^1/4''$ T-nut

$5''$

1 1—$3/4'' \times 16'' \times 20''$
1 2—$1^1/2'' \times 5'' \times 16''$
1 3—$3/4'' \times 1^1/2'' \times 5''$
1 4—$1'' \times 2^1/2'' \times 21''$

Table hole = $2^3/4''$ D.

②

①

Note: Use piece of hardwood
between lock bolt
and drill press
column

11-9 Plans for drum sander jig No. 2.

11-10 The drum sander jig can be used to drill holes on a common centerline.

Place the work on the table and press lightly against the drum. Remember that sandpaper works because its abrasive particles are actually cutting tools. Make passes steadily and without forcing. Excessive pressure will harm drum and work and will result in premature clogging of the abrasive.

Sanding regular or irregular curved edges is one of the highlights of drum sanding (FIG. 11-11). It's difficult to imagine another woodworking procedure that does the job as efficiently. Keep the work moving to prevent the drum from forming an indentation. Make passes *against* the rotation of the sander. If you do otherwise, the drum will take control of the operation away from you.

Remember that these are sanding jobs: the primary purpose is to smooth edges. If you must remove a lot of material, get to the final results by passing the work across the sander several times. Avoid working against one area of the abrasive. Occasionally, adjust the drum's vertical position so its entire length can be utilized.

JOINTING AND SURFACING

Jointing here does not refer to the kind of work that can be done on a jointer/planer, but rather a sanding operation that involves passing workpieces between the drum and a fence (FIG. 11-12). It's a practical way to form pieces to exact width, especially when many parts are needed. Don't try to remove a lot of material in a

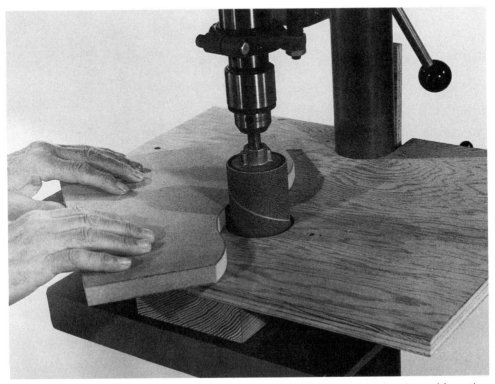

11-11 Workpieces are held flat on the table and pressed lightly against the drum. Move the work steadily in a direction that opposes the drum's rotation.

11-12 Edges can be jointed in this fashion. Make light cuts and keep the work moving from right to left.

single pass, and keep the work moving. If you hesitate during the pass, the drum will indent the work edge.

The same method can be used to smooth surfaces on material that is not wider than the drum (FIG. 11-13). It's an excellent way to get a fine finish on slats and other thin components.

INTERIOR EDGES

Sanding the edges of interior cutouts is just a matter of placing the work before adjusting the position of the drum.

PIVOT SANDING

Pivot sanding is a good technique to adopt when you want a disc of perfect diameter and with smooth edges. This does not differ from other pivot jig methods (FIG. 11-14). Because a drum sander jig does not provide too much frontal area, it will probably be necessary to add a platform for extra work support. The work is impaled on the pivot point, which can be simply a nail driven up through the platform, and then rotated in a clockwise direction. The radius of the work is controlled by the position of the platform.

11-13 Surfaces as well as edges can be smoothed by passing the work between the fence and drum. Pieces that are wider than the drum's length can be sanded if second pass is made with the work resting on its opposite edge.

11-14 The pivot jig technique can be used to produce discs with smooth edges and perfect diameters. The extra platform that is needed for the pivot point is clamped to the jig's table.

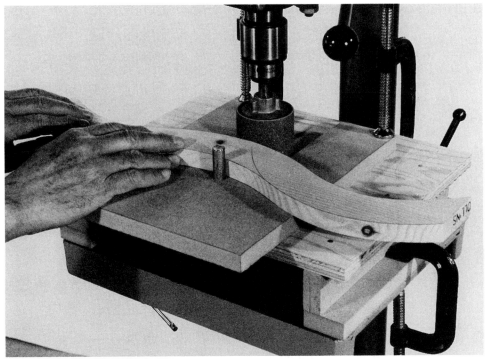

11-15 Components that are similarly shaped on opposite edges can be sanded to exact width by using this system. The work must be turned carefully throughout the pass to maintain its correct position between post and sander.

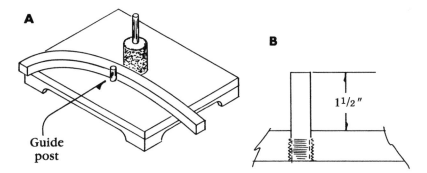

11-16 The guide post can be a dowel secured in a platform (A), or use a headless 1/2-inch bolt seated in a threaded sleeve (B).

SANDING PARALLEL CURVES

Components that have similar configurations on opposite edges can be sanded using the method shown in FIG. 11-15. The extra jig, which is tack-nailed or clamped to the sander's table, has a U-shaped cutout that provides for the drum and allows positioning of the guide post in relation to the width of the stock. The edge of the work that bears against the post must be smooth, so that phase of the operation is done freehand in normal fashion. For best results, feed the work so an imaginary, straight, lateral line between work and drum will always be tangent to both surfaces. Use this method only when you need several uniform pieces. Do the work without a guide when a single component is required.

The post can be a length of dowel secured in the platform. A post can also be made by removing the head from a $3/8$-inch or $1/2$-inch bolt and securing it with a threaded sleeve (FIG. 11-16).

Chapter **12**

Working with metal

There are similarities as well as differences between drilling in metal and drilling in wood. For one thing, while some metal-drilling tools can be used on wood, the reverse is rarely true. (Twist drills are an exception.) A good-size hole can be formed in wood in a single operation, but it's not prudent, and might not even be possible, to follow the same practice when drilling metal. The best procedure for optimum hole quality is to arrive at the required diameter by starting small and then changing to other bits to increase the hole size in stages. The amount of material each bit has to remove is minimized, as are the chances that a bit might snag in the hole. This applies to holes that are more than 1/8 inch or so in diameter.

SPEEDS AND FEEDS

The most efficient speed to use depends on the size of the hole and the density of the material. Soft metals, for example, can usually be drilled at higher speeds than steel. TABLE 12-1 offers reasonable suggestions for beginning speeds. As usual, there is the option of raising or lowering RPM if a change improves the operation and results in better output.

If the drill press won't produce the ideal speed, settle for the speed that comes closest. Stay on the low side for safety. The owner's manual will describe the machine's maximum drilling capabilities. If it advises, for example, not to drill more than 1/2 inch in steel, then follow these guidelines.

Use a feed pressure that will keep the bit cutting constantly. A timid feed that allows the bit merely to rub won't do the job, and will result in premature dulling of cutting edges. A good indication of correct speed, reasonable feed pressure,

**Table 12-1   Suggested Speeds (RPM)
When Drilling Non-Wood Materials**

HOLE SIZE	MATERIAL				
	Aluminum	Plastics	Cast Iron	Brass	Mild Steel
1/16″	4700	4700	4700	4700	2400
1/8″	4700	4700	2400	4700	1250
3/16″	4700	2400	2400	2400	1250
1/4″	4700	2400	1250	2400	700
5/16	1250	1250	1250	1250	700
3/8″	2400	1250	700	1250	700
7/16″	1250	1250	700	1250	
1/2″	1250	1250	700	700	
5/8″	700	700			
3/4″	700	700			

and a sharp bit, is normal waste that is escaping easily from the hole. The waste might be small particles or curled ribbons; it depends on the material (FIG. 12-1).

LAYOUT

Layout work on metal is usually done with a scriber (FIG. 12-2), although many workers use an awl that has a very sharp point. The problems with these tools are

12-1   Waste curling easily and smoothly from the hole is a sign that the bit is sharp and that feed and speed are correct.

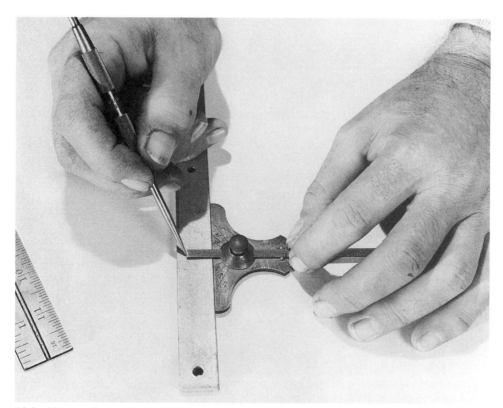

12-2   Use a scriber or a sharp awl for layout on metal. Pencil lines and dots would be too broad.

that a scribed line isn't too obvious on some metals, and it isn't good practice to incise lines that are difficult to remove.

Professionals solve the problem by using special surface coaters. These include materials like white and blue chalk, a mixture of white lead and turpentine, copper sulfate, and layout compounds. The coatings, when dry, show layout marks very clearly. Light scribe marks remove a tiny thread of the coating without harming the metal itself. Home craftsmen can coat the metal with a felt pen or spray or wipe it lightly with a quick-dry flat paint. Layout marks will be easy to see and the coatings can be removed with solvents (FIG. 12-3).

Hole locations should always be marked with a prick punch. For more accuracy, the prick punch mark can be enlarged by using a center punch (FIG. 12-4). The latter tool will form a conical indentation in which the point of the drill will seat when you start drilling. Without these precautions it's easy for the bit to wander off the mark. This is especially true when drilling radial holes into cylinders.

When the accuracy of the hole is extremely important use a pair of dividers to mark a hole-size circle around the prick punch mark. The circle acts as a guide so you can quickly tell if the bit is inclined to move off center. A possible remedy, should the bit try to wander, is to use a small punch to make several indentations on the side that is opposite of where the bit wants to go. The idea might bring the bit back on center.

12-3 Coating the material makes it easy to scribe fine lines that are clearly visible. Use the marking tool lightly to avoid incising the metal.

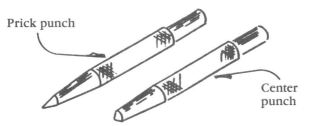

12-4 Mark hole location with a prick punch and then enlarge with a center punch. This will form an indent to start the drill bit and help prevent wandering.

WORK SECURITY

A bit that is forming a hole in metal develops a considerable amount of twisting action throughout the operation and especially when it breaks through the underside of the work. Take every precaution to avoid the hazard of a strip of metal free-wheeling on the end of a bit. Security can be provided by using clamps, some anti-twist arrangement, and special holding devices.

A simple but widely used arrangement that prevents hazardous work twist involves securing a pair of bolts in the table's slots (FIG. 12-5). The bolts are positioned so the hole location will be in line with the bit. Clamps, of course, can be

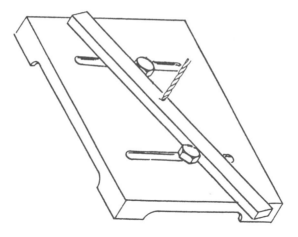

12-5 Twin bolts locked in the table slots will position the work and oppose the twisting action of the bit.

12-6 A drill press vise is standard equipment for metal working. Vs in its jaws make it easy to grip and to position round stock.

used, but the stop bolt method eliminates the nuisance of having to loosen and tighten a pair of clamps for each piece that needs drilling.

An excellent accessory for gripping workpieces is the drill press vise (FIG. 12-6). It is ideal for drill press work because the design allows easy clamping or

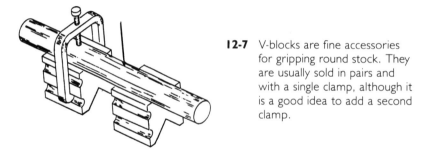

12-7 V-blocks are fine accessories for gripping round stock. They are usually sold in pairs and with a single clamp, although it is a good idea to add a second clamp.

bolting to the machine's table. The V-cuts in its jaws—sometimes horizontal, sometimes both horizontal and vertical—facilitate gripping round stock as well as flat pieces.

Flat or square work can rest at the bottom of the jaws, positioned so the drill, at breakthrough, will have clearance in the channel of the vise. When the size of the work is less than the depth of the jaws, place a wood block of equal thickness under it so the work will be raised for visibility and there will be a base for the bit's breakthrough.

V-blocks offer a practical means of gripping round stock (FIG. 12-7). They are usually sold in pairs and with a U-shaped clamp that has lips to engage the grooves in the blocks. It's wise to buy an extra clamp so workpieces can be secured to

12-8 V-blocks and a drill press vise make a good team for securing cylinders. In this concentric drilling operation, the Vs assure that the work is truly perpendicular.

both blocks. The V-blocks are often used with a drill press vise to grip round stock horizontally or vertically (FIG. 12-8).

Backups

You can achieve good work support by using a scrap block under the work as you do when drilling wood. Be sure that both the work and the backup are clamped to the table. Many workers use steel parallels for work support, but it's bad practice to place them too far apart (FIG. 12-9). The workpiece, especially if it is thin, will spring under feed pressure and cause the bit to jam, thereby increasing the twist hazard.

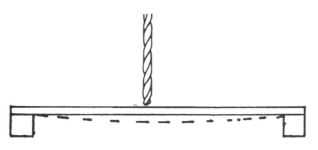

12-9 Parallels used under a workpiece must not be spaced far apart. This can cause the work to spring under feed pressure, increasing the likelihood that the bit will jam.

An accessory you can make that will be serviceable for many metal drilling chores is simply a V-block that can be clamped in place so the work will receive adequate support (FIG. 12-10). Overall, the essential purpose of a backup is to provide work support as close to the drilling area as possible.

12-10 Work support must always be as close to the drilling area as possible. A V-block serves the purpose very nicely. Use clamps to secure both items.

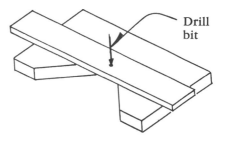

Drill bit

DRILLING THIN SHEETS

Sheet metal must have a generous wood backing that will keep the material flat as possible (FIG. 12-11). The work is taped to the backing and the pieces are held against a fence that, in addition to preventing twist, serves as a gauge when holes are needed on a common centerline. There is a tendency for very thin sheets to raise with the drill when it is retracted. Guard against this by retracting very slowly, or, if necessary, by using a hold-down to keep the work firmly in contact with the backing. To assure accuracy and eliminate the burring that occurs when

12-11 Sheet metal must have firm contact with a backing. Provide for this by using double-coated tape at midpoints.

drilling sheet metal, many operators sandwich the work between pieces of scrap wood.

Large holes can be formed in sheet metal by using a hole saw or fly cutter (refer to FIG. 2-16). It is critical to supply adequate support for the work and to use a clamping arrangement to prevent any work movement. The usual fly cutter precautions prevail: Use a slow speed and keep hands well away from the cut area.

Large, thin-walled tubing or duct work can be drilled by using the procedure shown in FIG. 12-12.

TAPPING

Tapping is the procedure used to thread holes for machine screws and bolts. One of the problems involved when doing the job by hand with a tap wrench is getting the cutter started squarely and keeping it that way throughout the operation. This is where the drill press can help. Remember, though, that the drill press merely assures squareness; the work is never done under power (FIG. 12-13).

Secure the tap in the chuck and turn the chuck by means of a lever that is inserted in one of the chuck key holes. Apply very light feed pressure as the tap is

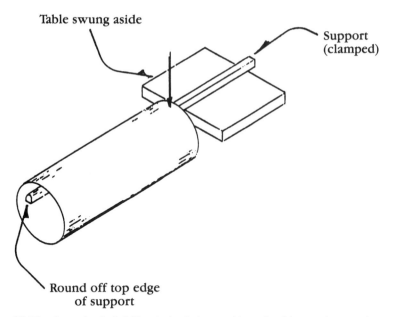

Table swung aside

Support
(clamped)

Round off top edge
of support

12-12 A method of drilling holes in large, thin-wall tubing or duct sections.

turned. Coordinate rotation and feed pressure so the tap will form threads without being forced. The method will be successful regardless of the type of tap that
is used (FIG. 12-14).

The tap functions by cutting metal. To remove waste material, rotate the
chuck counterclockwise about one quarter turn for every half turn made in the
cutting direction. The job will go easier if you apply a drop of light machine oil or
a special tapping oil to the cutter as you work. Use care when withdrawing the
tap. Remember it is captured in the hole and must be removed by rotating the
chuck counterclockwise, while maintaining some amount of feed pressure.

Tapping must be a precise operation in order for screws or bolts to turn and
to hold as they should. Therefore, the starting hole must be sized so the tap will
form "national coarse" or "national fine" threads that are exactly right for the
fastener. TABLE 12-2 lists the correct drill bit to use for various tap sizes.

COUNTERSINKING

Stove bolts and machine screws often have countersunk heads. They, like flat-
head wood screws, require a special indentation in the work so they will seat flush
with work surfaces. The cutter used is tempered to stand up to metal working and
is shaped to suit the head angle of metal fasteners. You'll get cleaner results if you
select a countersink whose net diameter is greater than the size of the screwhead.

The speed to use depends on the material. It's always best to start at a slow
speed and to increase it until the cutter is removing waste without a sign of chatter. When you have many countersinks to form, establish the depth of the first one
and then set the machine's stop rod so all the cuts will be the same (FIG. 12-15).

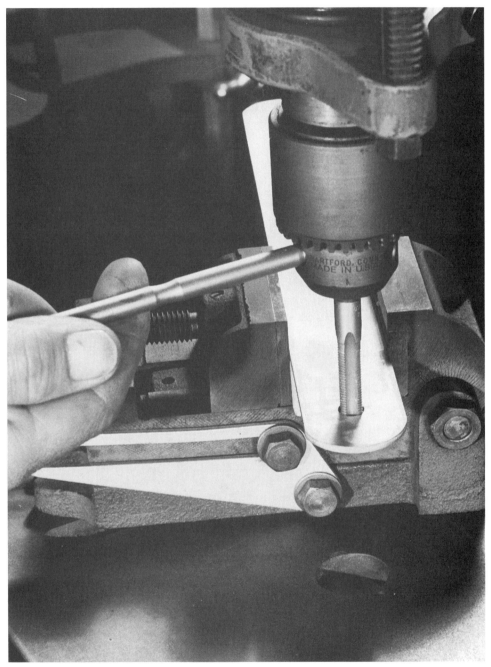

12-13   Tapping is a practical drill press application, but it is *not* done under power. The chuck is turned by hand as light feed pressure is applied.

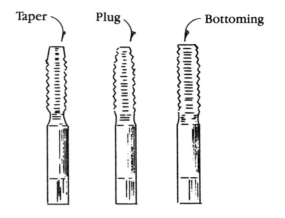

12-14 Three types of taps. All can be used successfully in a drill press.

Table 12-2   Tap Drill Sizes

SIZE OF TAP	NC OR NF	DRILL SIZE	DECIMAL EQUIVALENT
4-40	NC	43	.0890
6-32	NC	36	.1065
8-32	NC	29	.1360
10-32	NF	21	.1590
10-24	NC	25	.1495
12-24	NC	16	.1770
1/4-20	NC	7	.2010
1/4-28	NF	3	.2130
5/16-18	NC	17/64	.2656
5/16-24	NF	17/64	.2656
3/8-24	NF	21/64	.3281
3/8-16	NC	5/16	.3125
7/16-20	NF	25/64	.3906
7/16-14	NC	3/8	.3750
1/2-13	NC	27/64	.4129
1/2-20	NF	29/64	.4531
9/16-12	NC	31/64	.4844
9/16-18	NF	33/64	.5156
5/8-11	NC	17/32	.5312
5/8-18	NF	37/64	.5781
11/16-11	NC	19/32	.5937
11/16-16	NF	5/8	.6250
3/4-10	NC	21/32	.6562
34/-16	NF	11/16	.6875

NC = NATIONAL COARSE   NF = NATIONAL FINE

12-15 Countersinking is required for machine screws and stove bolts that have heads shaped like flat-head screws. Use the machine's stop rod so cuts will be uniform.

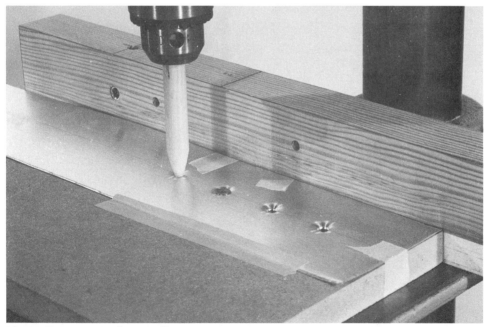

12-16 Sheet metal is "dimpled" to form the seat for flat-head fasteners. Normal countersinking won't work because it would enlarge the hole.

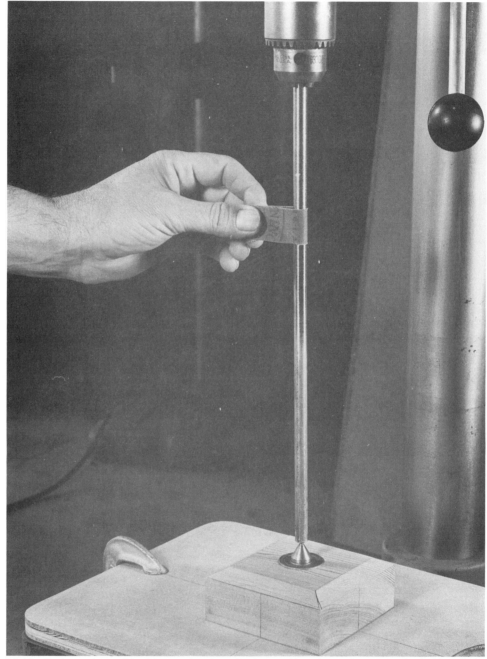

12-17 Rods can be polished by turning them in the machine. Long pieces must be held at their free end to avoid whipping.

DIMPLING

It is not a good idea to countersink sheet metal because the screw's hole will be greatly enlarged. The solution is to use the technique, called *dimpling*, illustrated in FIG. 12-16. First drill the screwholes through the metal and a backing. Remove the metal and countersink the backing as if it was going to receive the screw. Tape the metal back into position and then use a pointed hardwood dowel to form the dimple. Feed pressure alone might be enough to shape the metal. If not, use a slow speed and light pressure. A little dab of wax can be used as a lubricant.

POLISHING RODS

You can polish steel rods by gripping them in the chuck and sanding them with a piece of emery paper. If the work is long so that it might whip about under power, provide for gripping it at its free end. The jig used in FIG. 12-17 employs a lathe live center that is press-fitted in a wood block. The free end of the rod is indented enough so it can sit over the cone of the center. This is a rather fancy arrangement that was made for frequent use. For a one-time job, the rod can seat in a hole drilled in some scrap wood, that is then clamped to the table.

12-18 Spot polishing (damaskeening) provides a decorative finish on metal surfaces. It works best on soft metals.

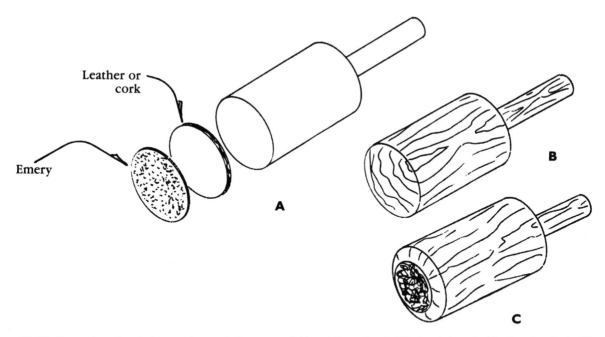

Leather or cork

Emery

A

B

C

12-19 Examples of tools that can be made for spot polishing: (A) steel rod drilled and threaded for headless bolt, (B) lathe-turned maple, (C) lathe-turned maple with cavity to hold steel wool.

SPOT POLISHING

Spot polishing or *damaskeening* is done to provide an attractive finish on metal surfaces (FIG. 12-18). Special tools can be made as shown in FIG. 12-19. The type of abrasive to use will depend on the hardness of the metal and whether you wish the spots to be coarse or fine. It's best to run some tests on scrap material before getting on with the actual work.

The most attractive results occur when the work is guided by a fence and the spots are uniformly overlapped. Feed pressure and drill speed have an effect on results, but you can judge the best way to go when doing the testing.

Whatever design you use for the tool, be sure its bearing end is flat and that it will be parallel to the surface of the work.

Chapter **13**

More ideas and applications

The more you use a drill press, the more you will realize how flexible a power tool it is. This chapter covers some more drill press applications and presents some practical accessories that you can make. Also covered are advanced uses for some of the basic projects covered in other sections.

MAKING A FOOT FEED

In many drill press chores it is necessary to frequently extend and retract the quill: routing, drilling a series of similar holes, and spot polishing are examples. A foot feed leaves both hands free to control the work and does away with the nuisance of having to lock the quill in extended position for each cut—a step that is required, for example, when routing is done on various areas of a workpiece, or when the same routing cut is needed on many parts.

The system is based on a treadle that is secured at the base of the machine (FIG. 13-1). Depressing the treadle pulls down a wire cable that passes over an intermediate pulley, and then a second one that is locked to the column under the headstock. The end of the cable is coiled around the hub of the feed mechanism and linked to one of the feed levers (FIGS. 13-2 and 13-3). On some machines it might be necessary to link the cable directly to a feed lever.

This idea is most feasible for a floor model unit, but it can be adapted to suit a bench machine if the tool is secured so its side is aligned with the edge of the bench. The treadle assembly can be mounted on a base that has a vertical member for the block on which the treadle bar pivots. If the machine is on its own stand, the cable can pass through a hole in the top of the stand. Check the dimensions against your machine before cutting components.

13-1 Treadle arrangement for a drill press foot feed. The spring is included so the treadle bar will return easily to its neutral position.

13-2 An intermediate pulley serves to keep the wire cable moving smoothly, and keeps the wire away from the table. The cable passes over a second pulley and links to a lever after being wrapped around the hub of the quill feed mechanism.

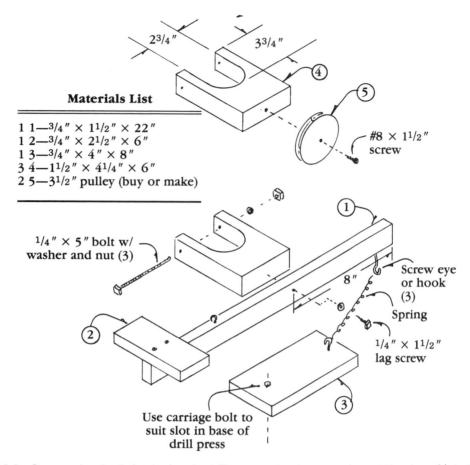

Materials List

1 1—³/4" × 1¹/2" × 22"
1 2—³/4" × 2¹/2" × 6"
1 3—³/4" × 4" × 8"
3 4—1¹/2" × 4¹/4" × 6"
2 5—3¹/2" pulley (buy or make)

2³/4"   3³/4"

④

⑤

#8 × 1¹/2" screw

①

1/4" × 5" bolt w/ washer and nut (3)

8"

Screw eye or hook (3)

Spring

②

1/4" × 1¹/2" lag screw

③

Use carriage bolt to suit slot in base of drill press

13-3   Construction details for the foot feed. The wire cable (not shown) should be about ¹/8 inch in diameter.

THE DRILL PRESS AS AN OVERARM ROUTER

The jig shown mounted on a drill press in FIG. 13-4, serves nicely for chores that ordinarily require a separate routing machine. An advantage of the idea over the drill press itself as a router, is that the portable tool provides higher speeds. Drill press routing, plus this additional accessory, should be enough equipment to cover any routing chore you might encounter, except those where work size makes it necessary to apply a portable tool by hand.

The prototype was designed around a portable router with a 6-inch-diameter base, and for a 15-inch drill press that has a 2³/4-inch diameter column. It might be necessary to alter some dimensions to suit the tools you will work with (FIGS. 13-5 and 13-6).

The best procedure is to make the top layer first, sizing the U-shape at one end to suit the drill press, and shaping the front end to conform with the router's base. Then make the bottom layer so it conforms with its mate, but with its front end designed to accommodate the router. Assemble the pieces with glue and No. 6-×-1-inch flat-head woodscrews.

13-4 A portable router is used with this jig to provide the features of an overarm pin router.

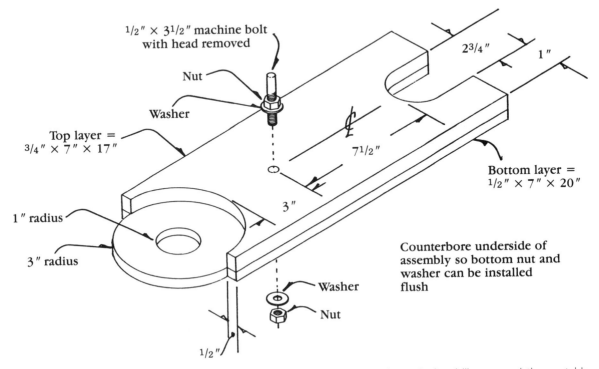

13-5 Plans for a portable router jig. The design might have to be altered to suit the drill press and the portable router that is used.

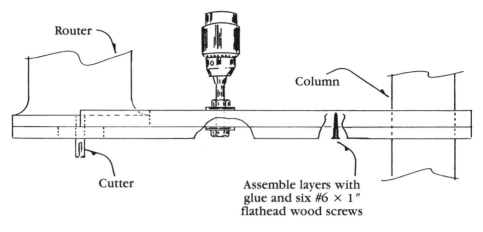

13-6 The jig secured in the drill press.

Place the assembly on the drill press table so it embraces the column, its center aligned with the spindle. Drill a pilot hole. Counterbore the underside so the washer and nut will be flush when tightened, then enlarge the pilot hole to 1/2-inch. Place the router's base in position and use it as a template to mark locations for attachment screws. The screws should match those that secure the tool's subbase and be long enough to pass through the bottom layer of the jig. Drive them upward to engage the threaded holes in the base. The router can be mounted with or without its subbase; doing without it will allow greater depth of cut.

In addition to the jig, the setup requires a large auxiliary table for work support (FIG. 13-7). This can be a sheet of 3/4-inch plywood that is bolted or clamped to the drill press table.

MORTISING JIG FOR ROUND STOCK

By forming square holes in round stock, it is possible to produce connections like those shown in FIG. 13-8. Adopt the idea for jobs like setting square or rectangular rails into round legs and to inset shelves into radial mortises formed in posts.

Conventional mortising tools are used—that is, quill-mounted casting and chisel/bit combinations. These must operate in conjunction with special work-holding devices (FIG. 13-9). The jig shown here uses a regular mortising hold-down, but if you follow the construction details in FIG. 13-10, you can make one of hardwood.

Operational considerations like speed and feed and squareness of chisel don't change. Be certain the vertical centerlines of spindle and V are common and that you move work so that successive cuts will follow the same path. It's best, in addition to marking the ends of the mortise, to draw parallel lines to define cut width (FIG. 13-11).

When you need the same cut on many pieces, it's a simple matter to tack nail a strip of wood so it spans the V to serve as a stop. It's also the way to go when forming radial mortises. The stop will assure that the cuts will be on the same line (FIG. 13-12).

13-7 An auxiliary table added for work support. A fence, which should not be too high, can be tack-nailed in place. If the jig moves during cutting, use a large clamp at the back end to tighten the U-shaped cutout against the column.

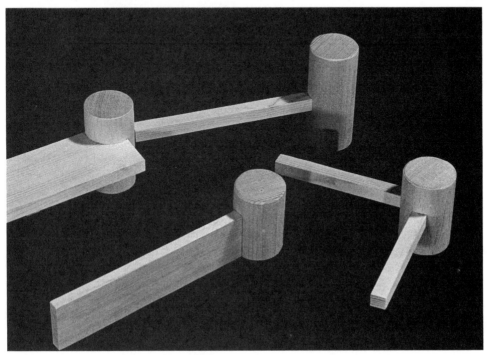

13-8 Typical jobs that can be accomplished when a jig is made for mortising round stock.

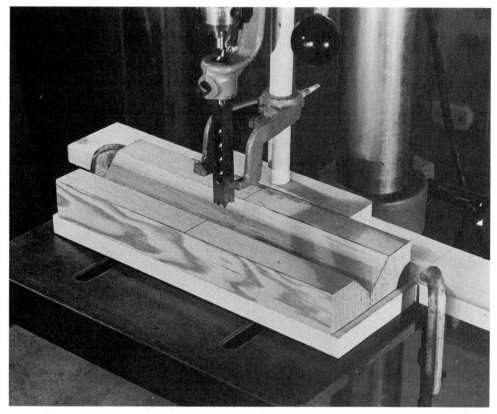

13-9 The jig used with standard mortising tools. Even the commercial hold-down can be used if the jig is designed this way.

The same arrangement is used to form mortises on the corners of square stock (FIG. 13-13). Feed very carefully when making first contact with the work.

MAKING MOLDINGS BY DRILLING

You can create decorative pieces to use as a molding by drilling holes that follow a particular pattern or by forming holes on the joint line of pieces that are clamped together (FIG. 13-14). The parts are separated after drilling and strip cut on a table saw to produce individual thin or thick strips that can be joined together or mounted individually on a backing (FIG. 13-15).

The same idea can be utilized to make cornice-type components with scalloped edges. In this case, a hole saw of fly cutter is used to form large holes on the centerline of board that is wide enough for two projects. Making a saw cut on the center of the stock is the final step (FIG. 13-16).

Strip-cut pieces can be joined so half circles conform, or with the cuts off-set so they produce a completely different effect (FIGS. 13-17 and 13-18). The parts can be joined with glue to form a flat panel or they can be mounted on a flexible backing so they can turn corners or form a circle. The backing can be self-adhesive felt or strips of tape, or you can glue them onto canvas.

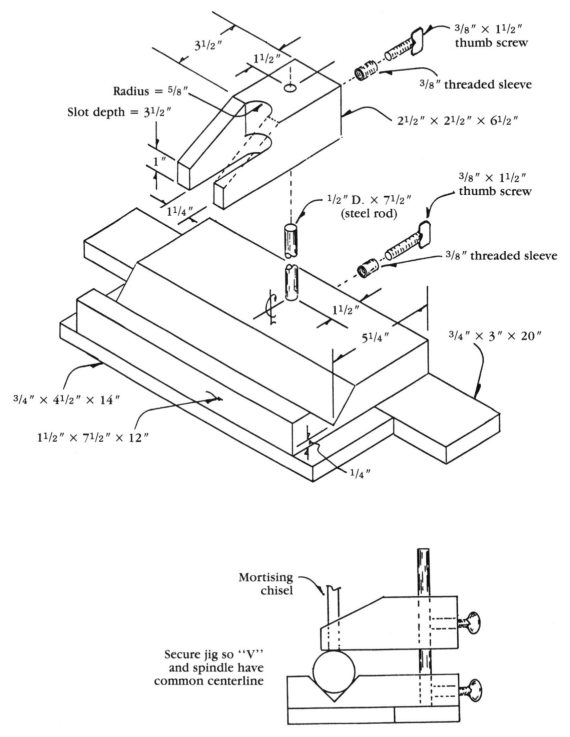

3½"

1½"

3/8" × 1½"
thumb screw

Radius = 5/8"

3/8" threaded sleeve

Slot depth = 3½"

2½" × 2½" × 6½"

1"

3/8" × 1½"
thumb screw

1¼"

½" D. × 7½"
(steel rod)

3/8" threaded sleeve

1½"

3/4" × 3" × 20"

5¼"

3/4" × 4½" × 14"

1½" × 7½" × 12"

1/4"

Mortising
chisel

Secure jig so "V"
and spindle have
common centerline

13-10 Plans for the special mortising jig. The design includes a hold-down made of hardwood.

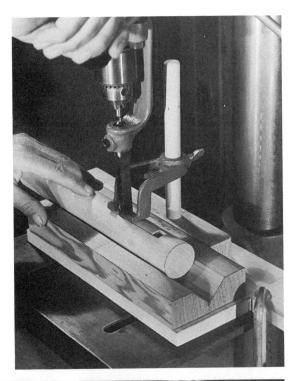

13-11 The jig is positioned so the V shape and the spindle will have a common, vertical centerline. The cutting is done in routine fashion; make end cuts first, then clean out waste with additional, overlapping cuts.

13-12 A strip of wood spanning the V serves as a stop. Radial mortising is done by rotating the work for each cut. Here too, make end cuts first.

13-13 Mortising can be done into corners of square stock, but make initial contact very carefully so the cutters won't wander off the mark.

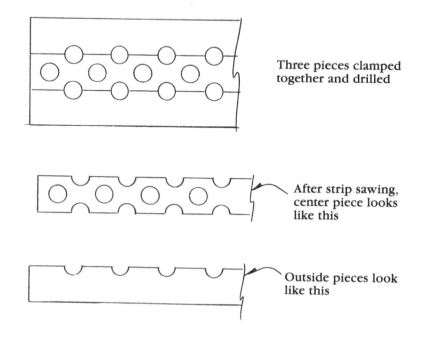

Three pieces clamped together and drilled

After strip sawing, center piece looks like this

Outside pieces look like this

13-14 Producing moldings by drilling.

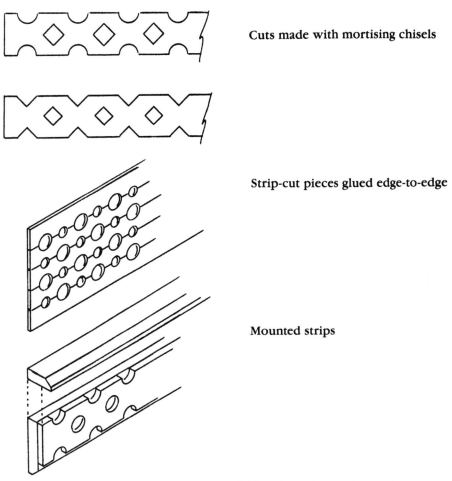

Cuts made with mortising chisels

Strip-cut pieces glued edge-to-edge

Mounted strips

13-15 Strip cutting the parent stock produces individual pieces that can be used in various ways.

Mortising chisels, for square or diamond shapes instead of holes, can be used in similar fashion (FIGS. 13-19, 13-20, and 13-21).

SURFACE CARVING

You can do surface carving with a fly cutter if you limit the depth of the cut (FIG. 13-22). A preliminary step should be planning the layout on paper so locations for the tool's pilot drill can be established. Actually, you can work more accurately if you drill pilot holes before working with the fly cutter. In addition to decorative surface cuts, this idea can be used to embellish components like escutcheons for drawer pulls, locks, and so on (FIG. 13-23).

If you have the type of fly cutter that works with a sloping bit instead of a vertical cutter, you can create interesting angles and patterns by overlapping cuts (FIG. 13-24). The pilot holes that are formed by the cutter can be left open or they can be plugged with dowels or buttons.

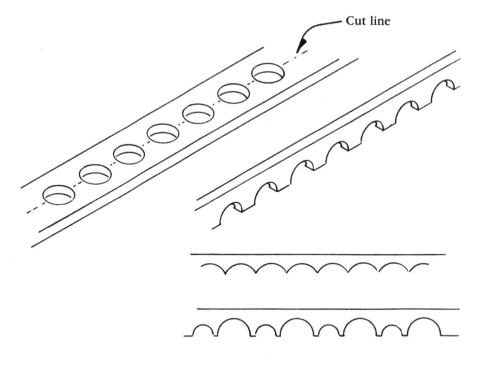

Variations

13-16   Produce components with scalloped edges by forming large holes down the center of a board and then sawing the board in two.

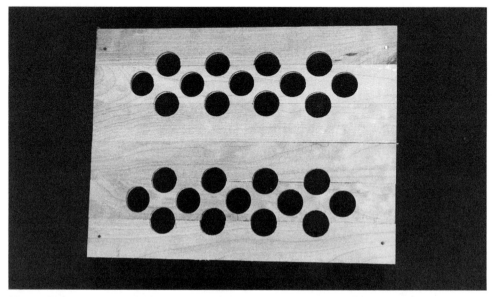

13-17   Strip-cut pieces can be joined edge-to-edge to form panels. Here, parts were assembled so the holes conform.

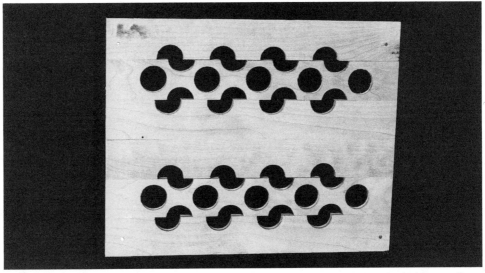

13-18 A completely different effect results when the strips are assembled so the holes are offset.

13-19 Cuts formed with a mortising chisel are square, but look diamond-shaped when the cutter is turned 45 degrees.

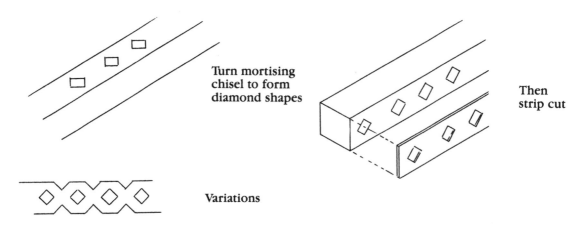

Turn mortising chisel to form diamond shapes

Then strip cut

Variations

13-20 Examples of cutting that can be done with a mortising chisel.

13-21 Strip cutting is done on a table saw. Use a pusher when doing this kind of work so your hands can be well away from the cut area.

13-22 Surface carving with a fly cutter. Some of the segments that result from the cutting can be removed later with a chisel.

FORMING RINGS

A double-blade fly cutter is a good tool for forming rings because its bits are individually adjustable. One bit is set for the ring's outside diameter, the second bit cuts on the inside diameter (FIG. 13-25). You can form rings with a one-bit fly cutter by adopting the following procedure.

Nail the workpiece to its backing close to the pilot drill. Form the outside diameter. Remove the waste and then set the bit for the ring's inside diameter. Be careful at the end of the second cut because the ring will be free. Remember: The nails driven into the backing are the only means of keeping the workpiece secure during the final cut.

PATTERN SANDING

The jig for pattern sanding consists of a clamped auxiliary table that secures a disc so it is concentrically aligned with a drum sander. The disc, which exactly matches the diameter of the drum, serves as a guide for a pattern on which workpieces are tack nailed or secured with double-coated tape (FIG. 13-26). The pattern, whose thickness can't be greater than the thickness of the guide, is a duplicate of

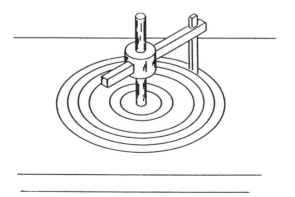

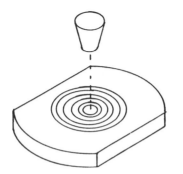

13-23 A fly cutter can be used to form or to decorate components like escutcheons for drawer pulls or switch plates.

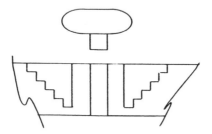

the shape that is needed. Saw workpieces close to approximate form. When they are moved against the drum, they will be sanded to match the pattern.

The problem with the basic system is that sanding is done on a small area of the drum and it can clog or become worn pretty quickly. The solution is to have a hole for the drum through the table and to use a ring instead of a disc as the guide (FIG. 13-27). Then, because the drum can be moved vertically, its entire abrasive surface is utilized. When a ring guide is used, the pattern must be reduced to compensate for the cross section width of the ring (FIG. 13-28). The basic drum sander table can be adapted for pattern sanding by adding a guide ring. For temporary use, this can be tack-nailed in place (FIG. 13-29).

13-24 A fly cutter with a sloping blade produces some interesting effects when cuts are overlapped. The pilot holes can be left open or sealed with plugs or buttons.

13-25 A double-blade fly cutter provides a convenient way to produce rings with a single cut.

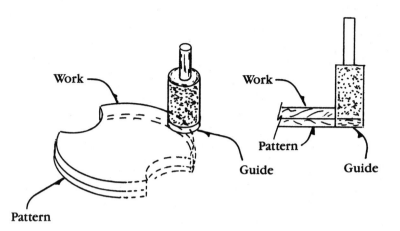

13-26   The basics of pattern sanding. The thickness of the pattern must not be more than the thickness of the guide.

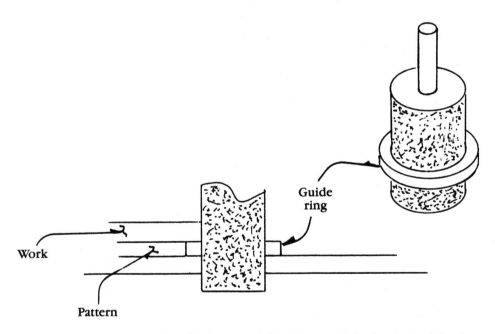

13-27   Using a ring instead of a solid piece as a guide allows vertical adjustment of the drum. Thus, the entire abrasive surface can be utilized.

PATTERN ROUTING

The technique for pattern routing is similar to pattern sanding. For routing, the guide is a short post that is secured in an auxiliary table and aligned vertically with the router bit. The pattern is tack-nailed to the underside of the work and then placed over the post. One edge of the pattern must be firmly against the post when you bring down the router bit for depth-of-cut. Then, move the work while

13-28 When a ring guide is used, the pattern must be reduced by the cross-section width of the ring.

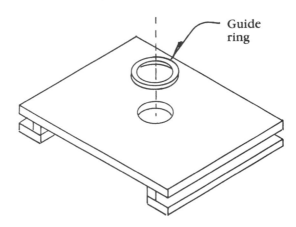

13-29 A ring guide can be added temporarily or permanently to a basic drum sander table.

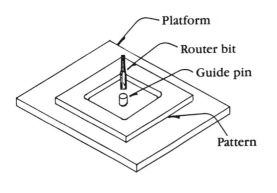

Platform
Router bit
Guide pin
Pattern

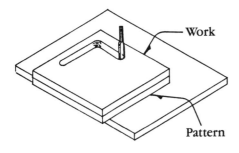

Work
Pattern

13-30 The setup for pattern routing. The pattern, with work attached, is moved so it bears constantly against the guide post.

13-31 The setup required for doing a limited amount of lathe work on a drill press.

maintaining constant contact between post and pattern (FIG. 13-30). This isn't difficult to do because you soon develop the ability to "see" the guide even though it's hidden by the workpiece.

THE DRILL PRESS AS A LATHE

The drill press can't be substituted for an individual lathe, but it will serve nicely for shaping turnings of reasonable length. Assuming that the chuck is a headstock, you must then add a tailstock to secure the free end of the work and a tool rest for bracing cutting tools (FIG. 13-31). The tailstock can be the simple affair (FIG. 13-32), or it can be more elaborate (FIG. 13-33). The advantage of the ball-bearing design is that the bolt will turn with the work (like the live center in a lathe), thereby eliminating the wear factor that exists when a spindle is turning on a fixed point. If the first idea is used, check quill extension occasionally to be sure the tailstock end of the work hasn't loosened.

The tool rest, constructed like the one in FIG. 13-34, is clamped to the table so its support edge is 1/8-inch or so behind the vertical centerline of the work. When doing the turning, it's better to work with a scraping rather than a cutting action. Woodworking files can be used with good effect; final finishing is accomplished with strips of abrasive (FIGS. 13-35 and 13-36).

You can produce mini turnings on the drill press by working on cylinders with diameters no greater than the chuck's capacity. The shaping is done with various files. Because the files will tend to clog rather quickly, keep a file card close by and use it frequently to clear the file's teeth. This is a convenient way to make project components like rail posts for small shelves, finials, and pulls for small drawers.

SANDING TRICKS

There are various ways to make drum sanders of particular diameter and length (FIG. 13-37). Dowels are a good source of material, and special cylinders can be made by turning. Hole saws or fly cutters will produce discs with edges smooth enough to take sandpaper.

The drums can be mounted on a bolt or a shaft can be provided by using a headless lag screw or a hanger bolt. When the drum is small enough, it is simply gripped in the chuck (FIG. 13-38).

Long drums should be installed along the lines suggested for drill press turning. Actually, if you have or plan to do some turning, the same arrangement can be used to mount a large drum sander (FIGS. 13-39 and 13-40).

The easiest way to provide the abrasive surface is to use self-adhesive sandpaper. A spray adhesive can also be used to attach the sandpaper. Read the directions on the container for application methods and safe use. If you would like a resilient sanding surface, cover the drum with thin rubber sheeting or cork before applying the abrasive.

Flap sanders are handy for smoothing contoured edges and for jobs, such as enlarging a hole, that might be a bit tight for a fastener or another component. Making them is a simple procedure. Cut fine slots in the end of the cylinder so they will hold strips of sandpaper securely. The end of the strips can be left whole

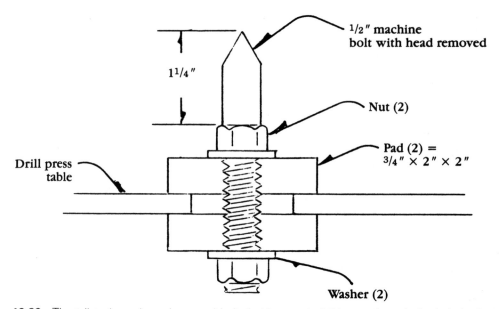

1/2" machine bolt with head removed

1 1/4"

Nut (2)

Pad (2) = 3/4" × 2" × 2"

Drill press table

Washer (2)

13-32   The tailstock can be a sharpened bolt that is mounted this way through the hole in the drill press table.

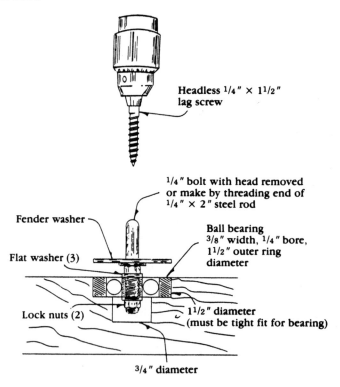

Headless 1/4" × 1 1/2" lag screw

1/4" bolt with head removed or make by threading end of 1/4" × 2" steel rod

Fender washer

Flat washer (3)

Ball bearing 3/8" width, 1/4" bore, 1 1/2" outer ring diameter

Lock nuts (2)

1 1/2" diameter (must be tight fit for bearing)

3/4" diameter

13-33   A tailstock that rotates with the workpiece eliminates the wear that occurs when a spindle turns on a fixed point. The drive center can be a headless lag screw or a hanger bolt.

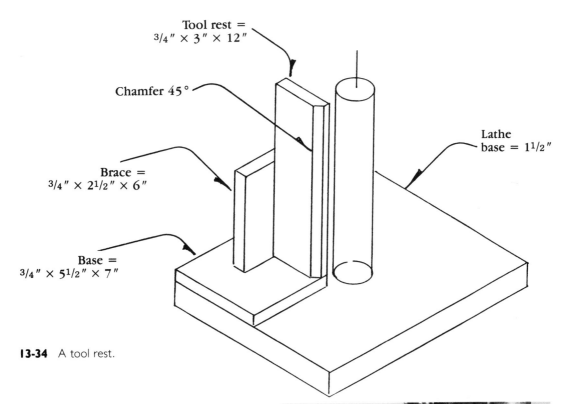

Tool rest =
3/4″ × 3″ × 12″

Chamfer 45°

Lathe
base = 1¹/₂″

Brace =
3/4″ × 2¹/₂″ × 6″

Base =
3/4″ × 5¹/₂″ × 7″

13-34 A tool rest.

13-35 Files can also be used for forming details. Keep the file cutting teeth clean by using a card file.

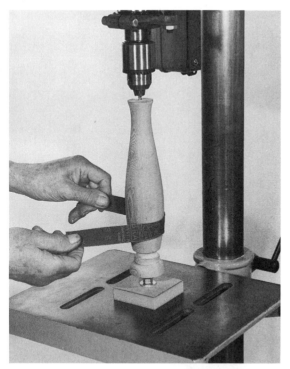

13-36 Use strips of sandpaper to do the final smoothing. Keep speeds under 2000 RPM, and don't try to shape pieces that are more than about 3 inches in diameter or more than about 12 inches long.

13-37 Types of drum sanders you can make by using round stock and by cutting discs with fly cutters and hole saws.

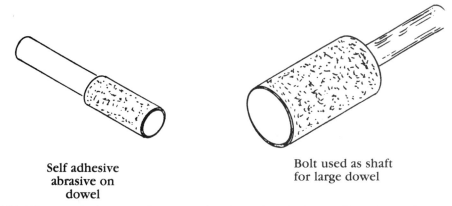

**Self adhesive
abrasive on
dowel**

**Bolt used as shaft
for large dowel**

13-38 Homemade drums can be mounted on bolts. Dowels that are 1/2 inch or under can be gripped directly in the three-jaw chuck.

13-39 Long sanding drums must be secured at their free end. The setup is similar to the system used for lathe work.

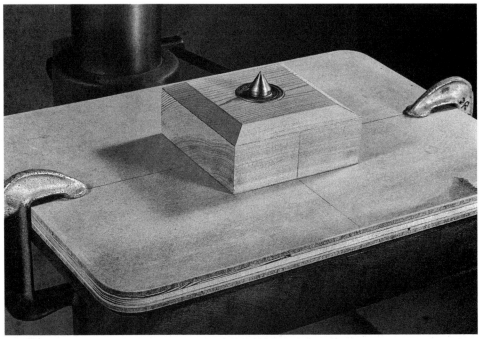

13-40   A live center can be press fitted in a wood block and used to provide support at the free end of the drum sander.

13-41   Flap sanders are easy to make. Just saw thin slots in the end of cylinders so strips of sandpaper can be held securely.

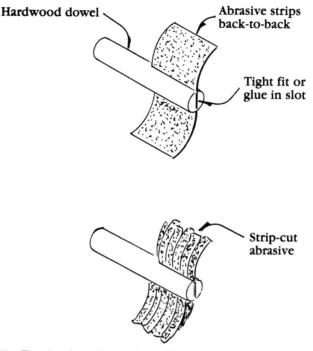

13-42   The abrasive strips can be solid or their ends can be strip-cut.

or they can be strip-cut to conform more closely with contoured edges or surfaces (FIGS. 13-41 and 13-42).

Dowels can be tapered by using a jig like the one shown in FIG. 13-43. The dowel is gripped in the chuck and the jig is positioned before the machine is turned on. This is a good way for ship model builders to form masts and spars.

FLUTING

The jig detailed in FIG. 13-44 makes drill press fluting a feasible procedure. The jig is sized to suit the length and diameter of the project, and it provides nail pivots on which the work can be rotated. A lock nail at one end, driven into the work, keeps it from turning when cutting is done. Holes for the lock nail should be spaced on a circle so that the work can be rotated a certain number of degrees for each cut.

Use the setup shown in FIG. 13-45 when doing the fluting. The two fences will assure that the work is moved in a straight line. Be sure the jig is placed so work and cutter will have a common, vertical centerline. If the drill press table can't supply enough work support, set up an auxiliary platform before organizing the jig.

SURFACING

A rotary planer can be used to true or smooth surfaces or to reduce the thickness of stock. The tool, called a Safe-T-Planer, is often used in a conventional chuck,

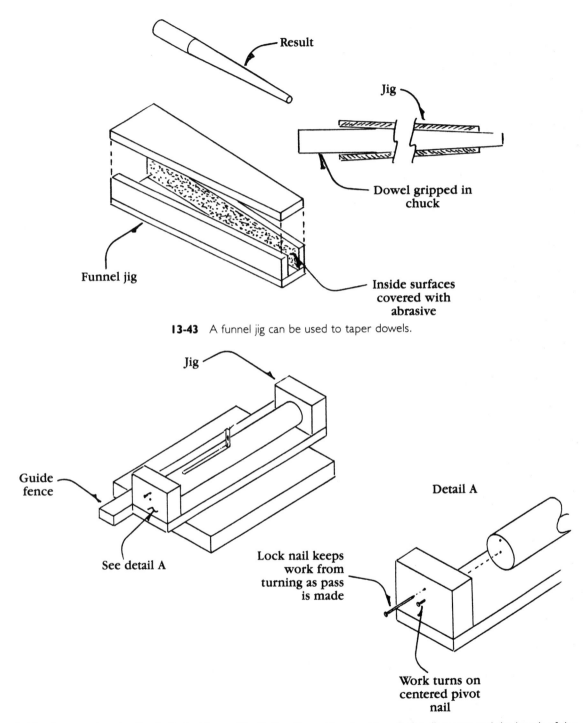

Result

Jig

Dowel gripped in chuck

Funnel jig

Inside surfaces covered with abrasive

13-43 A funnel jig can be used to taper dowels.

Jig

Guide fence

See detail A

Detail A

Lock nail keeps work from turning as pass is made

Work turns on centered pivot nail

13-44 Construction details of a jig that is used for fluting. It must be sized to suit the diameter and the length of the workpiece.

13-45 Double fences make it easy to guide the jig on a straight line. A router bit is used to form the flutes. Use a high speed and make repeat passes if necessary.

but can also be mounted in a router chuck. Surfacing is done by providing a fence to support and guide the work, and by making the pass from left to right (FIG. 13-46). Covering more area than the tool can accomplish in a single pass is just a matter of moving the fence. The planer is not designed for deep cuts, but this doesn't matter because best results are obtained by making light cuts and repeating them if necessary.

Forming wide rabbets is another job that can be accomplished with the planer (FIG. 13-47). The fence is just a length of 1½-inch stock with a central, semicircular cutout that accommodates the tool (FIG. 13-48). The same setup can be used to form wide tongues. Simply make a second pass after the stock has been inverted and turned end-for-end.

DRILLING NON-WOOD MATERIAL
Glass

Glass drilling is not difficult to do on a drill press if proper procedures are followed. Many workers use a short piece of brass tubing as the drill. The tubing, with an outside diameter that suits the hole to be drilled, is slotted at one end with

I3-46 Surfacing can be done with a Safe-T-Planer. Use a fence for work control; make passes from left to right. Several light cuts are better than a single deep one.

I3-47 The rotary planer can be used to form wide rabbets. Use the same setup to form tongues.

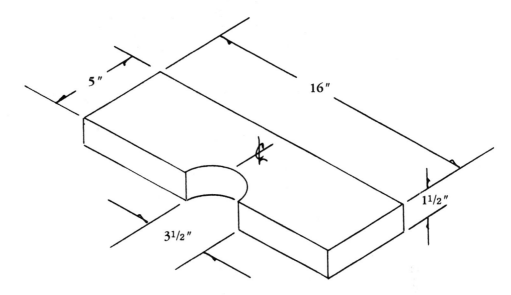

13-48 This fence comes in handy when the rotary planer is used for rabbeting.

a fine saw. The slots (there can be more than one) don't have to be more than about 1/4-inch long.

The first step is to use putty to form a dam around the hole area. The purpose of the dam is to contain a mixture of silicon-carbide abrasive powder and a light oil, or even a bit of turpentine. The combination of tubing and abrasive powder forms the hole by grinding. Use the lowest spindle speed and keep feed pressure to a minimum, only enough to assure that the bit is grinding. When possible, complete the hole by grinding from both sides of the material. This will avoid cracking that might happen when the tube exits from the work.

Another system employs bits that are designed for glass drilling (FIG. 13-49). They can be used dry but there is no harm in using them with the dam/grinding powder technique. In all cases, be sure the glass is supported by a perfectly flat backup. A precaution against cracking is to rest the work on a piece of felt or some indoor-outdoor carpeting (FIG. 13-50).

Ceramics

The special bits for glass drilling can be used to form holes in ceramic tiles (FIG. 13-51). Here too, it's essential that the work be perfectly flat and firmly in place. Otherwise, feed pressure can cause the material to crack. Conventional carbide-tipped masonry drills are also good tools for tile drilling (FIG. 13-52). Use slow speeds and take it easy with feed pressure. On the other hand, use enough pressure or the bit won't do much cutting.

Plastics

The special glass drilling bits do a fine job forming clean holes through plastics (FIG. 13-53). Ordinary twist drills may be used but they tend to dig in, especially on

13-49 These special bits are fine for drilling glass and some other hard materials.

13-50 A putty dam that holds a fluid mixture of abrasive powder and light oil makes drilling jobs go more smoothly. Use a slow speed and keep feed pressure at a minimum.

13-51 The special glass bits also work on ceramic tile. Be sure work and backing are absolutely flat.

13-52 Conventional carbide-tipped masonry bits also do a good job on ceramics. Placing the work on a resilient pad is a good idea.

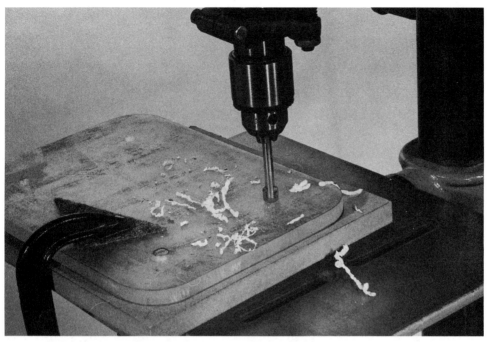

13-53 What you must guard against when drilling plastics is excessive heat build-up. This is especially important when the material has thermosetting characteristics.

13-54 It's easy to drill clean holes through paper if the material is held tightly between top and bottom covers. Do the drilling with brad point bits.

softer plastics. Retract bits frequently to avoid pile-ups of waste and to keep bit and material as cool as possible. Be especially careful when drilling thermosetting plastics. Heat can cause the material to soften or burn.

Paper

Paper products can be drilled cleanly if sandwiched tightly between scrap pieces of wood (FIG. 13-54). Use a speed between 1000 and 1200 RPM and work with the brad point bits. Retract the bit frequently because cutting forms rings of paper that will cling to the end of the cutter.

Glossary

backup   Scrap wood placed between work and table to prevent splintering when the bit breaks through.

base   Bottom support for the machine. Designed for securing the floor. Sometimes used as a table.

belt guard   Covers drive mechanisms: belts and pulleys.

belt tension   Tautness of V-belts. About 1/8 inch to 1/4 inch of slack is usually recommended.

belt tension lock   Ensures correct belt tension after adjustment.

bench model   A machine with short column for mounting on a stand or workbench.

bevel lock   Keeps table secure after it has been tilted for angular drilling.

bevel scale   Guide for setting table to a particular angle.

blind hole   A hole that stops short of going through the workpiece. Also called a stopped hole.

chuck   The device that secures bits and other cutters. Usually a three-jaw chuck.

chuck key   Used to tighten the chuck so cutting tools will be gripped securely.

column   The vertical tubular support for all drill press components.

column support collar   Locks to column to keep headstock at correct height.

counterbore   A limited depth hole, usually made so plugs can be used to hide screws or other fasteners.

countersink   An inverted cone shape, usually made so flathead screws can be driven flush with surfaces.

depth stop lock   A nut or other device that is used to limit quill extension.

depth stop rod   Part of depth stop mechanism. Moves vertically with the quill.

depth scale   Graduations on the depth stop rod that predetermine quill extension.

drill bits Cutters for forming holes. Most common bits are twist drills, brad point bits, spade bits, and Forstner bits.

drill vise A clamping device for work security and operator safety. Most often used when drilling metal.

drum sander A cylindrical abrasive tool for smoothing curved edges. Available in various diameters.

flute Spiral-shaped grooves on bits that provide a route so waste chips can escape from a hole.

fly cutter An adjustable cutter for forming large holes.

head lock Secures the headstock in correct position on the column.

headstock Heavy casting that encases most of the tool's working mechanism.

hold-down Any device, commercial or homemade, that secures work in correct position and adds to safety.

hole saw Cylindrical devices with saw teeth on perimeters. Used for forming large holes or discs.

jig A homemade device that extends a tool's scope or makes it easier to work more accurately or safely.

key (chuck) Used to tighten the chuck so cutting tools will be gripped securely.

mortising A drill press function for forming square holes. The application requires special accessories.

plug cutter Forms wooden cylinders for use in counterbored holes to hide screws. Some products are long enough to form tenons and short dowels.

quill Encases the spindle. Quill and spindle move together vertically.

quill feed levers Used to move the quill downward. May be a single lever or a trio of radial handles.

quill lock Will secure the quill in an extended position. Commonly required when doing work like routing or shaping.

quill return An adjustable spring device that automatically returns the quill to a neutral position.

quill travel The maximum amount that a quill can be extended. This determines maximum hole depth that can be formed normally.

router chuck A special gripped device that is used in place of the regular chuck when using router bits.

shaping A drill press function that allows some applications normally done on an individual shaping machine. Requires special accessories.

spindle The spindle, encased in the quill, is the rotating component. In a quality drill press, it is mounted on heavy-duty ball bearings.

spindle adapters Special holders for such work as shaping and routing that are used in place of the regular chuck.

spindle speed   Supplied by step pulleys or a variable speed mechanism. Spindle speed is the RPM of the cutting tool.

table   The surface on which workpieces are secured.

table lock   Secures the table at a selected height.

table support   Casting that connects the table to the column.

table tilt   Adjustment of table for angular drilling.

through hole   A hole that goes completely through the workpiece.

variable speed   A mechanism that provides infinitely variable speeds from minimum to maximum, as opposed to step pulleys that provide specific speeds.

work feed   How the work is moved. Applied mostly to chores like routing and shaping. Feed pressure is the force with which the quill is extended.

Index